How Literature Plays with the Brain

How Literature Plays with the Brain

The Neuroscience of Reading and Art

Paul B. Armstrong

Johns Hopkins University Press
Baltimore

© 2013 Johns Hopkins University Press
All rights reserved. Published 2013
Printed in the United States of America on acid-free paper

Johns Hopkins Paperback edition, 2014
9 8 7 6 5 4 3 2 1

Johns Hopkins University Press
2715 North Charles Street
Baltimore, Maryland 21218-4363
www.press.jhu.edu

The Library of Congress has cataloged the hardcover edition of this book as follows:

Armstrong, Paul B., 1949–
    How literature plays with the brain : the neuroscience of reading and art /
Paul B. Armstrong.
        pages   cm
    Includes bibliographical references and index.
    ISBN 978-1-4214-1002-9 (hardcover : alk. paper)—ISBN 978-1-4214-1003-6
(electronic) — ISBN 1-4214-1002-8 (hardcover : alk. paper) — ISBN 1-4214-
1003-6 (electronic)
    1. Reading, Psychology of.   2. Psychology and literature.
3. Neurosciences and the arts.   4. Literature—Psychology.   I. Title.
    BF456.R2A86 2013
    801'.92—dc23

                                                        2012044603

A catalog record for this book is available from the British Library.

ISBN-13: 978-1-4214-1576-5
ISBN-10: 1-4214-1576-3

*Special discounts are available for bulk purchases of this book. For more
information, please contact Special Sales at 410-516-6936 or specialsales@
press.jhu.edu.*

Johns Hopkins University Press uses environmentally friendly book
materials, including recycled text paper that is composed of at least 30
percent post-consumer waste, whenever possible.

For Beverly

Contents

Preface   *ix*

1   The Brain and Aesthetic Experience   1

2   How the Brain Learns to Read and the Play of Harmony
    and Dissonance   26

3   The Neuroscience of the Hermeneutic Circle   54

4   The Temporality of Reading and the Decentered Brain   91

5   The Social Brain and the Paradox of the Alter Ego   131

    Epilogue   175

    Notes   *183*
    Index   *213*

Preface

Why should a professor of literature like myself find neuroscience so fascinating? That is a question I have often asked as I found myself swept away by a growing sense of excitement and urgency while reading the technical, often dauntingly difficult neurobiological literature about action potentials, neuronal assemblies, phase-locking oscillations, mirror neurons, and so forth. Why immerse myself in this tough stuff when I could be reading a novel? At the risk of sounding naive and trite, I attribute this fascination in part simply to my interest as a humanist in what makes us human. That is an interest that neuroscience and literary studies share. Literature matters to me, among other reasons, for what it reveals about human experience, and the very different perspective of neuroscience on how the brain works is part of that story. What has surprised and excited me, however, is the plethora of unexpected convergences I have come across between the experimental findings of neuroscience and what I know as a literary critic and theorist about reading, interpretation, and the aesthetic experience. Again and again, while reading neuroscientific accounts of the structure and functions of the brain, I have been struck by how these matched up with views I had developed from a lifetime of thinking, teaching, and writing about the experience of reading and the interpretation of literary texts. These similarities are extensive and deep, I think, for reasons that have to do with fundamental brain processes at play in the aesthetic experience. Working out these parallels and convergences in detail is the primary purpose of this book.

My central argument is that literature plays with the brain through experiences of harmony and dissonance that set in motion and help to negotiate oppositions that are fundamental to the neurobiology of mental functioning—basic tensions in the operation of the brain between the drive for pattern, synthesis, and constancy versus the need for flexibility, adaptability, and openness to change. The brain's ability to play in a to-and-fro manner between competing imperatives and mutually exclusive possibilities is a consequence

of its structure as a decentered, parallel-processing network consisting of reciprocal top-down, bottom-up connections among its interacting parts. Experiences of harmony and dissonance of the sort typically associated with
art facilitate the brain's ability to form and dissolve assemblies of neurons,
establishing the patterns that through repeated firing become our habitual
ways of engaging the world, while also combating their tendency to rigidify
and promoting the possibility of new cortical connections.

The claim that art is associated with play, harmony, and dissonance is not
surprising, of course. As I explain in the first chapter, there is a long tradition
going back at least to Kant and continuing up to the present day that views
play as integral to the aesthetic experience. I also show there that an opposition between viewing harmony as the distinguishing feature of art and viewing dissonance as such is pervasive in the history of aesthetics. This is not
accidental, I argue, given the centrality of play, harmony, and dissonance to
the functioning of the brain. What would be disconcerting would be to find
that the way the brain works did not match up with the reports that readers,
critics, and theorists have offered over the years about what happens when
they experience literature and art. That these accounts keep returning to play
as a central feature of the aesthetic experience—while diverging drastically
in the sorts of to-and-fro interactions they find promoted by art (culminat-
ing in unity, synthesis, and balance or provoking defamiliarization through
disruption and transgression)—is a fact about human experience to which
the history of aesthetics testifies, and it is a fact that neuroscientific accounts
of the brain help to explain. We have the kind of brain that thrives by playing
with harmony and dissonance, and the experiences that have so widely and
typically been reported about encounters with art and literature are correlated in interesting ways with basic neuronal and cortical processes.

How these processes relate to particular aspects of reading and interpretation is what is surprising and (to my mind) fascinating. For example, it turns
out that the age-old claim of hermeneutics, the theory of interpretation, that
understanding is inherently circular finds confirmation in the neuroscience
of reading and vision. The contemporary hermeneutic theory that describes
reading as a process of gap filling and consistency building is corroborated by
the findings of neuroscience on how vision constructs shape and color and
on how the brain "recycles" (Stanislas Dehaene's useful term) biologically
inherited functions for invariant object recognition into the capacity to read
graphic signs. The ability of readers and other interpreters to find conflicting

meanings in the same state of affairs has roots in what Semir Zeki calls "the neurobiology of ambiguity." The experience of surprise and frustrated expectations so characteristic of aesthetic experience is based on the temporality of coupling and decoupling cell assemblies that Francisco Varela describes as the neural correlate of the experience of the durational moment. If we have aesthetic emotions that paradoxically both are and are not the real thing, that is related to what Antonio Damasio calls the "as-if body loop," whereby the brain fools the body into having vicarious experiences. This is one of the ways in which neuronal structures (including the ability of some brain cells to mirror observed actions, as Giacomo Rizzolatti discovered) make possible the paradoxical experiences we have of other people as both intersubjectively connected to us through a world we share and solipsistically isolated from us in minds we can at most simulate or theorize about—a paradox that literature both enacts and overcomes when it allows us to temporarily live other lives from the inside. These are some of the convergences between neuroscience and aesthetics that I find compelling and that I try to work out in the following chapters.

It is important to be clear from the outset, however (as I explain in chapter 1), that explicating these convergences is not an attempt to reduce art to neuroscience or to solve what is known as the "hard problem" of how the electrochemistry of brain cells and the interactions of neuronal assemblies give rise to consciousness and aesthetic experience. To discover correlations between neurobiological processes and reports about experience is not to claim causality. Nor is it to privilege the perspective of neuroscience (or of aesthetics) as the final answer about the phenomenon in question. Each perspective has its uniquely defining strengths (and limitations), and one should no more want to privilege either to the exclusion of the other than one would want to use only one kind of instrument to solve every problem one came across. There are things that aesthetics and literary theory can tell us about the experience of reading that the perspective of neuroscience cannot, and the reverse is also true. This difference constitutes what philosophers call the "explanatory gap" that divides these fields, but it is my contention (and a basic premise of this book) that this gap also provides an opportunity for exchange across it whereby each side can benefit from what the other has to say from its unique vantage point.

My ability to move back and forth across this gap has to do not only with my willingness to learn enough neuroscience to become a competent interlocutor with the scientific community but also with my knowledge of phenomenology, a philosophical tradition of reflection about the structure of consciousness, experience, and interpretation that in recent years has been seen by a growing number of philosophers and scientists to have important connections to neuroscience. One of the problems for neuroscience is its need for rigorous, reliable descriptions of experience against which to check its experimental findings at the neuronal level and from which to draw hypotheses about the neural correlates of consciousness. Rather than rely on personal intuition or so-called folk psychology, some neuroscientists have found it useful to consult the subtle, detailed descriptions of such phenomenologists as Edmund Husserl and Maurice Merleau-Ponty about our lived, embodied experience of the world. A figure common to both neuroscience and phenomenology is the nineteenth-century American psychologist and philosopher William James, whose pioneering *Principles of Psychology* stands up very well and is frequently cited in the neuroscience literature and whose strand of pragmatism is widely seen as a precursor of phenomenology. (My doctoral dissertation and first published book, long ago, on the connections between William James, his novelist brother Henry James, and phenomenology's theory of experience was, looking back, a first step in my progress toward this current book.)

The "neurophenomenologists" have had much to say to neuroscience about cognition and the intentionality of consciousness, about the horizonal structure of temporality, and about the embodied experience of self and other. So far, however, they have drawn little, if at all, from phenomenology's extensive contributions to the theory of interpretation, reading, and aesthetics. And that gives me my opening. These are areas of prime importance for neuroaesthetics, areas about which I can claim to speak with some authority from my work on phenomenological hermeneutics and literary theory.

Many of the topics I consider have been of concern to the so-called new cognitive literary studies, which has received a good deal of attention in the national press. As will be evident from the criticisms of this field that I offer along the way, I have mixed feelings about it. I was initially excited about its contribution to the long-overdue return of interest in the reading process and in aesthetic questions that has recently occurred in literary studies (although at least some of the cognitive critics unfortunately seem to be disavowing this interest by attempting to rebrand their work as "cognitive cultural studies"). I have been disappointed, however, by the absence (with a few notable exceptions) of serious engagement with neurobiology. Most of the cognitive critics focus on psychology and its studies of the mind rather than on neuroscience and its analysis of the brain. This is understandable, perhaps, because the translations from psychology to literature are easier to make than those from neurobiology to art, but it is also regrettable because of the controversies that exist (as I explain in chapter 1) between cognitive psychology and neuroscience. There are important overlaps and connections between the experimental findings of these two fields, and in an ideal world each would be a resource for the other (as they sometimes are), but the "mind-brain" divide is a problem that cognitive literary studies needs to address (and that I think neurophenomenology can provide a way past).

Another serious shortcoming of cognitive literary studies, at least in its current state, is its neglect of phenomenology and hermeneutics, even though their long and rich traditions provide a trove of conceptual tools for analyzing the relations between aesthetic phenomena and the processes of interpretation, cognition, and meaning creation (whether these are attributed to the mind or to the brain). The reasons for this neglect may have to do with the desire of this new field to distance itself from the reader-response criticism of the 1970s, when some kinds of literary phenomenology enjoyed brief prominence, which has since fallen into deep disfavor. Whatever the explanation, this deficiency is unfortunate because (as I try to show) phenomenology and hermeneutics provide a variety of useful bridges to connect aesthetics and neurobiology across the explanatory gap dividing these fields.

I hope this book will be read by neuroscientists as well as literary critics. For the neuroscientific community, it offers suggestions about how the findings of many different areas of research—from the neurobiology of vision and reading to the temporality of brain waves, the brain-body interactions underlying emotions, and the role of mirror neurons in imitation, learning, and self-other relations—may be connected to a variety of widely attested aesthetic and literary phenomena. In the absence of a single, unified definition of art and the aesthetic, neuroaesthetics stands to benefit from the guidance of literary theory about what matters and what doesn't (for example, don't try to find a single center of the brain for aesthetic experience—there isn't one, for both neuroscientific and aesthetic reasons—but do tell us more about how the brain's own reciprocal, top-down, bottom-up interactions relate to aesthetic effects of harmony and dissonance).

For critics and students of literature, I hope this book contributes to a return I see happening on a variety of fronts to questions basic to the humanities: What is aesthetic experience? What happens when we read a literary work? How does the interpretation of literature relate to other ways of knowing? Rather than offering new ways of reading and thinking about literature, this book tries to explicate the neurobiological bases of aesthetic experiences and processes of interpretation that are much in evidence and widely testified to across differences in history, methodology, and literary preference. Explaining how these differences—between the aesthetics of harmony and dissonance, for example, or between conflicting methods of interpretation—relate to basic brain processes is also one of the book's aims. Another important finding is that some aspects of the aesthetic experience that seem suspiciously fuzzy or even mystical to many tough-minded critics—for example, how reading entails filling in gaps left by the text or how literary works seem to be inhabited by a consciousness that comes alive when we read—turn out to have material foundations in the biology of brain processes. When literature plays with the brain, many wonderful things can happen that we typically associate with the aesthetic experience. One of the marvels of this experience is that it is based on biological processes that neuroscience can help us understand.

No one writes alone, and I could not have written this book without the help of friends, colleagues, and students whose assistance I am glad to acknowledge. (The ways in which this book falls short, however, are my responsibility alone.) I have always learned as much from my students as I have taught them, and this book in particular owes much to conversations about neurophenomenology with Jen Hall, whose own work in neuroaesthetics I have had the privilege to supervise through the Institute for Doctoral Studies in the Visual Arts. George Smith, the visionary founder of IDSVA, also gave me formative advice about an early draft. As my ideas were first developing, I imposed on the generosity of Jim Phelan to ask for feedback about the initial chapters, and his encouragement and candid, constructive criticism were invaluable. I benefitted immensely as well from the responses of Jeremy Hawthorn, Steve Mailloux, and Jeff McCarthy to an early version of the book. Ulla Haselstein, Don Wehrs, Andrzej Pawelec, Ann Kaplan, Marty Hoffman, and Vanessa Ryan also provided important criticisms and advice at various stages. Johns Hopkins's anonymous literary referee made some exceptionally insightful suggestions that guided a final round of reshaping and rewriting and significantly improved the book.

It has been a top priority for me to get the science right, and I was gratified when the anonymous neurobiologist to whom Hopkins also sent the manuscript gave it a thumbs up. This could not have happened without the generous assistance I received from friends in the scientific community who carefully read earlier drafts and patiently corrected my mistakes. I am very grateful to the neurobiologist Gary Matthews and the cognitive scientist Richard Gerrig, both former Stony Brook colleagues, and to Jim McIlwain, a Brown neuroscientist (and scholar of medieval literature), for the detailed criticisms and suggestions they gave me. Any errors that remain are evidence of my inability to take full advantage of their tutelage. Nor are they to blame for the aesthetic arguments and speculations that I develop based on the science.

This is the first book of mine that my children have found interesting and have actually contributed to in substantive ways. My daughter Maggie, a RISD-trained printmaker, did the technical work of preparing the book's figures for publication. She also gave me the benefit of a practicing artist's response to my arguments. The opportunity to collaborate with her as a peer was an unanticipated pleasure of this project. Her brother Tim, a Stanford-trained econometrician, was more bemused and skeptical about his humanist father's forays into neuroscience, but he offered good counsel about how the social sciences have tried to apply the findings of cognitive science. My teenage son Jack, still living at home, had to suffer more than his older siblings from my eagerness to share my findings ("Please, Dad, not another fact about the brain!"), but explaining my ideas to him and justifying their importance was perhaps the most rigorous test they had to pass.

Beverly, my partner in all things, was a more willing and sympathetic but no less critical audience. I dedicate this book to her, my favorite interlocutor.

How Literature Plays with the Brain

CHAPTER 1

The Brain and Aesthetic Experience

There are many good reasons to undertake cross-disciplinary studies, but one compelling justification is that a problem one wants to solve cannot be adequately addressed with the tools of one's discipline alone. That is clearly the case with neuroscience and art. If neuroscientists want to understand how the brain responds to art, music, or literature, they need to consult experts in those fields not only to appreciate the subtleties and complexities of these phenomena but also to avoid common traps, oversimplifications, and dead ends that serious theoretical study of aesthetics would disclose. A scientist might feel tempted to avoid this detour and go straight to the aesthetic experience on the grounds that everyone has an intuitive sense of art and beauty. That is no doubt what the neuroscientist V. S. Ramachandran felt when, after several days of strolling among Indian monuments, "one afternoon, in a whimsical mood, I sat at the entrance of the temple and jotted down what I thought might be the 'eight universal laws of aesthetics,' analogous to the Buddha's eightfold path to wisdom and enlightenment. (I later came up with an additional ninth law—so there, Buddha!)"[1] To be fair, the jocular, ever-bemused Ramachandran may realize how preposterously presumptuous this (literally) back-of-the-envelope philosophizing might seem (so there, Kant and Aristotle!), and some of his "universal laws" do indeed have merit, especially his insights into the role of distortion in artistic representation.[2] But no neuroscientist would take seriously a humanist who examined his or her own mind one day and proposed eight (or nine) universal laws of brain functioning on the grounds that we all have brains, just as we all experience art. In order to understand a complex phenomenon like the aesthetic experience, neuroscience obviously needs guidance from the humanities about exactly what it should be trying to make sense of.

It is perhaps less obvious what the humanities need from neuroscience.

Some claims that humanists make about the cultural and historical relativity or the universality of phenomena like art, language, and perception need to be tested against what neurobiology has discovered about the plasticity, as opposed to the inherent structural limitations and fixed characteristics, of the brain. As I explain in chapter 2, the *neuronal recycling* through which reading emerged provides an interesting, complex commentary on this question and shows that while the brain is not a completely blank slate, it can be written on (and even rewrite itself) in a surprising variety of ways.[3] But knowing how the brain reads and how literature plays with our neuronal processes will not necessarily lead anyone to change the way he or she interprets a text or alter the kinds of pleasure and instruction he or she receives from aesthetic experiences. The point of this book is not to persuade anyone to read differently but rather to explore how the functions and structures of the brain are correlated with what readers have widely reported about literature and aesthetic experience across history, from a variety of critical perspectives, with a range of conflicting, sometimes mutually exclusive presuppositions and interests about language, literature, and life. To understand the neuronal processes underlying the consistency building and gap filling that occurs when we read, for example, is not to privilege or exclude any particular practice of reading or any one set of literary values. If a neuroaesthetic study were to claim that some respected, widely shared views about the experience of reading literature were simply wrong and did not match up with the way the brain works, that would call into question the science for refusing to consider and account for evidence that did not agree with it.

The purpose of the examples from literature and literary theory that I offer in the following pages is to illustrate and document this variety of responses and the heterogeneity of literature. One of the distinguishing features of literary criticism and literary art (although it is not unique to them) is their variability—the conflict of interpretations that is characteristic of humanistic inquiry, the historical and cultural contingency of artistic value, and the range of sometimes opposing phenomena that have been associated with aesthetic experience. The challenge for neuroaesthetics, I contend, is to account for this variety—for the ability of readers to find incommensurable meanings in the same text, for example, or to take pleasure in art that is either harmonious or dissonant, symmetrical or distorted, unified or discontinuous and disruptive. Neuroscience cannot tell us how to settle these disagreements (and it shouldn't try), but its model of the brain as a "bushy," decentered assembly of recursive, parallel-processing operations does have much to say about how and why literature can play with the brain in such an extraordinary variety of ways. The brain is a peculiar, at times paradoxical, but eminently functional combination of constancy and flexibility, stability and openness to change, fixed constraints and plasticity, and these contradictory, paradoxical qualities are reflected in the workings of literature and literary interpretation in ways that can (I will argue) mutually illuminate the neurobiology of the brain and the experience of art.

For reasons that deserve a little reflection here at the outset, it is not likely that neuroscience will provide us with new interpretations of literary texts or tell us why we should prefer one reading rather than another. Neuroscience cannot tell us whether there are actually ghosts in *The Turn of the Screw*, for example, or the governess is simply hallucinating, but its experimental studies of the perception of multi-stable figures can help to explain how the brain responds to this kind of ambiguity. The main usefulness of neuroscience for literary studies is not as a source of new readings, and the following pages do not include extensive interpretations of particular literary texts. Some of the phenomenological theories I describe about how texts set up and play with readers' horizons of expectation are amenable to such treatment, as I and other phenomenological critics have shown elsewhere at length, but that is not the aim of this book.[4] Rather, my intent is to offer a range of examples from literature and literary theory to demonstrate the variability of the aesthetic phenomena that neuroscience must account for, and focusing in depth on a few individual works would not serve this purpose.

If a literary critic takes inspiration from neurobiological models or cognitive science to interpret a particular text, his or her hypotheses will enact the recursive processes of hermeneutic circularity that I describe in chapter 3. But it is also the case, for that reason, that one should not expect neuroscience to become a machine for producing readings. Although some critics have drawn on the findings of neurobiology to offer interpretations of literary texts, there is good reason for the skepticism voiced by Norman Holland, the literary theorist perhaps most knowledgeable about contemporary neuroscience: "I do not think neurology will be able to say useful things about particular texts or particular readings of particular texts for the foreseeable future. All we know of the brain now is some fairly broad, even universal systems."[5] This is not just a consequence of the generality of current neurobiological models, however, or the crudity of its measuring instruments.

Interpreters must generate hypotheses anew each time they encounter a text (even if their guesses about meaning may follow particular patterns from habits they have developed over a lifetime of reading), and science cannot take the place of hermeneutic intuition. Critics who draw on neuroscience or cognitive psychology for hunches to guide their literary interpretations must still engage in the to-and-fro of interpretive hypothesis testing.

Understanding the neuroscience of reading will not necessarily make us better interpreters. Neurobiological insight does not translate directly into improved practice in any field. No one would expect, for example, that an analysis of the neurobiological processes of vision and motor control at work when a baseball player hits a home run would improve performance at the plate (it may indeed have the reverse effect by making the hitter self-conscious about processes that need to happen beneath notice to function well). Neuroscience can, however, deepen our appreciation of the seemingly miraculous but materially grounded processes that occur when we read and interpret—or hit a home run, an act that seems to defy the biological limits of the temporality of vision and motor control given the milliseconds required to form a cross-modal neuronal assembly and the speed of a fastball heading toward the plate. My wonder about the neurobiological processes underlying reading, literature, and interpretation (also baseball) has increased since I've started studying neuroscience, even if my own abilities in these areas have not improved.

The technological revolution in brain imaging has opened astonishing new horizons for understanding mental processes, including what happens when we read, but it is also important to remember what technology cannot do. As I explain in chapter 2, functional magnetic resonance imaging (fMRI) technology has identified in the lower left occipital region of the brain a *visual word form area* (VWFA), which plays a central role both in the visual recognition of the constancy of objects and in the deciphering of graphic signs, and this discovery tells us much about the repurposing of visual processes that makes reading possible. But this technology is still too crude to identify exactly what is happening there when we read a particular text, let alone when we read different texts or when various readers construe the meaning of the same text differently. Too many neurons are covered by each *voxel* in the image, and the scanning process takes too long (because of the change in blood flow that it measures) for the spatial and temporal resolution of fMRI technology to offer anywhere near the necessary precision.[6]

This technology has been used by neuroscientists to compare the activity in different brains when viewers see particular films, for example, and these experiments have revealed perhaps surprising commonalities in the cortical areas activated in different viewers by the same movie. The Princeton neuroscientist Uri Hasson has shown that there is greater correlation in cortical activity in response to a Hitchcock film, curiously, than to a spaghetti western by Sergio Leone, a Larry David comedy, or a video of random pedestrian activity in a city square.[7] Although Hitchcock was notorious for wishing to control his viewers' response (remember the shower scene in *Psycho*), Hasson acknowledges that it is unclear what to make of the lesser degree of correlation shown with other films. It may not mean that Hitchcock is the greater artist, because some avant-garde directors who leave more open for the viewer have also produced important films by following a different aesthetic, and some very controlling films are propaganda (e.g., Leni Riefenstahl's powerful but for that reason frightening, morally disturbing Nazi documentary *Triumph of the Will*). The cortical activity of viewers may be correlated because of great artistic control or the intent to indoctrinate. Conversely, a lower degree of correlation may signal a movie's artistic success in inspiring a variety of imaginative, critical responses, or it may be evidence that the viewer is bored, distracted, or daydreaming. Furthermore, even when there is high correlation, fMRI images cannot tell what is happening in the neuronal assemblies of viewers who may disagree about how to interpret a particular film (as passionate Hitchcock critics often do). The imaging technology alone cannot make these discriminations.

We are far from the day when imaging technology would allow us to track with temporal and topographical precision the moment-to-moment unfolding of the neurobiological processes set in motion by aesthetic experiences. What is more, important aesthetic questions about the meaning of the phenomena mapped by these instruments will never be settled by the technology alone. Even if and when we develop the capability to track these processes down to the cellular level with millisecond accuracy, the "hard problem" (as it is called) of how consciousness and lived experience emerge from chemical and electrical processes at the cellular level will not be solved. The miracle of how neurobiology is transformed into conscious experiences of art and literature will remain. As the philosopher Colin McGinn puts it, "The water of the physical brain is turned into the wine of consciousness, but we draw a total blank on the nature of this conversion. . . . There is something terminal about our perplexity."[8] As in other areas of human (and probably also nonhuman) experience, pointing to where the brain lights up in a scan or charting complex interactions between cortical regions may not give a fully adequate account of art's lived actuality. Here as elsewhere, there is reason to believe that what philosophers call the "qualia" of consciousness escape full neurological mapping.[9] As the neuroscientist Adam Zeman explains, "A certain kind of subjective, first-person knowledge seems to lie beyond the reach of science."[10]

The neuroscientist Semir Zeki, who has done pioneering work in aesthetics, is probably right that "if you can tell the difference" between kinds of experience, "it is because different brain areas, or cells, are involved."[11] If someone has an aesthetic experience, after all, it must involve neuronal and cortical processes of some kind. But that does not mean that mapping this activity or analyzing its physiology is the best or most adequate way to explain it. There are cortical processes involved whenever a carpenter hammers a nail, or a lawyer makes an argument in court, or a baseball player steals a base, but the neurology of these events is not a satisfying or even appropriate account of a house, justice, or America's pastime. Although a state of affairs like carpentry, the law, or baseball may be linked to neurological activity, they all entail more than that activity alone can explain.[12] There is a distinctiveness to each domain that resists epistemological reduction and that requires modes of explanation that respect the integrity of its aims, purposes, and values. The same is true of art.

Nevertheless, any explanation of something translates it into terms other than itself, and the dangers of reductionism will not be avoided by refusing to compare different domains. It is possible to make comparisons that do not attempt to make one domain subordinate to the other and that remain mindful of what is gained as well as lost in translation. Experiments about brain functioning in aesthetic situations are invariably linked to lived experiences of art, and literary theory has things to say about those experiences that neuroscientists ignore at the risk of committing rudimentary fallacies that a good university course on the history of criticism would expose. As the neuroscientist Martin Skov wisely advises, "Experimental aesthetics cannot get started without using the insights from traditional aesthetics."[13] John Hyman similarly warns that "if we ignore the philosophy of the past, we shall simply reinvent the wheel" and "our ideas will be based on mediocre and amateurish philosophy of our own."[14]

The hard problem may be intractable, but it still leaves much to talk about—more, probably, than would be the case if it could be simply and conclusively solved. Another way of thinking about the hard problem is to see that it is responsible for what some neurophilosophers call the "explanatory gap" that exists between distinctive, irreducible levels of analysis: "Although neuroscientists have supplied neural models of various aspects of consciousness, and have uncovered evidence about the neural correlates of consciousness (or NCCs), there nonetheless remains an 'explanatory gap' in our understanding of how to relate neurobiological and phenomenological features of consciousness."[15] Neuroscientists and literary theorists of reading may not be able to close this gap, but they can still usefully talk across it. In everyday communication, after all, it is not only what we share but also what separates us that provides the basis for interaction; otherwise there would be no reason to exchange ideas, opinions, and perspectives. As the phenomenological theorist Wolfgang Iser points out, "social communication" requires not only commonalities—shared interests and the ability to translate one's language into another's—but also "asymmetry"—gaps, disjunctions, and differences that make "dyadic interaction" possible and provide its impetus, energy, and purposes.[16] Similarly, the explanatory gap dividing neurobiological processes and aesthetic experiences is a structure that enables exchange from positions of distinctive disciplinary difference without worrying about whether these conversations will ever overcome it.

That is why it is useful and necessary to go back and forth between different discourses (as I try to do throughout this book), ever again crossing this divide, in order to illuminate the parallels and correlations between neurological processes and aesthetic experiences. These processes and experiences are connected but distinctive and not finally reducible to one another. The different vocabularies of neuroscience and aesthetics can clarify and explore these differences, but we should not expect them to resolve them. Neuroscience and aesthetics see the states of affairs they study through different "terministic screens" (to borrow Kenneth Burke's useful phrase), vocabularies and codes that direct the attention in particular ways while deflecting it from others.[17] Different terministic screens are typically unable to perceive their defining ratio of insight and blindness by themselves, within the constraints of their own perspectives. That ratio can emerge, however, when one screen is juxtaposed against another, with a different configuration of attention and deflection, and this is one reason why the sometimes frustrating disagreements between incommensurable perspectives can nevertheless be illuminating even when they are not resolved. The disciplinary divide (or explanatory gap) across which neuroscience and aesthetics can talk with each other—but without ever eradicating their defining differences—is a case of conflict between terministic screens, and such conversation need not be talk at cross-purposes if we recognize the limits as well as the uses of what exchanges of this kind can hope to accomplish.

The sciences are commonly thought to be more in touch with "reality" and more certain of their knowledge than the humanities. But anyone who has studied the neuroscience literature knows that some areas of inquiry are more settled than others. The neural anatomy of vision is well established and unlikely to be overturned, but much remains unclear about brain rhythms and the synchronies that link neuronal assemblies. The details and scope of mirror-neuron functions also are still subject to lively dispute, as I explain in chapter 5, even if the evidence is very strong that a variety of imitative processes are at work across the cortex. All of these findings are a result of very indirect laboratory procedures that are a testimony to the imagination of the scientists and the ingenuity they show not only in the development of their measuring machinery but also in the design of their experiments. To study neurobiology is consequently to appreciate what a creative activity science is. It is not the case, however, as some crude versions of so-called social constructionism suggest, that science is simply a set of historically contingent fabrications produced and validated by a community of investigators.

To take the oft-cited case of Boyle's air pump, which created a vacuum in laboratory conditions, ask the bird that died in one of his experiments if the artificiality of the instrument means its results aren't real. "Yes, the facts are indeed constructed in the new installation of the laboratory and through the artificial intermediary of the air pump," the anthropologist of science Bruno Latour explains, but, he observes, "facts" that are produced by "actors" (scientists) in a "network" of relations (the measuring equipment and the scientific community) are not for that reason fictions or illusions.[18] Not all theories are equally provisional, as the misinformation campaigns about evolution and climate change remind us.

The areas of neuroscience where the "facts" are well established should cause humanists to take notice and revise their views accordingly (for example, as I will show, about the historicity and universality of certain processes, like reading). And when humanists invoke science, they should try to do so accurately and avoid flights of fancy about "thermodynamic capitalism," "the metaphysics of biology," or "the affectivity of matter"—irresponsible language that one too often finds in speculative but uninformed writing about literature and science, which confirms scientists' prejudices about the methodological laxness of the humanities.[19]

But it is not a rap against the epistemological rigor of the humanities that interpreters may (and typically do) disagree about how to construe the meaning of a novel or a poem. There is a difference, for one thing, between the hermeneutic conflicts that may divide interpreters with opposing assumptions about the purposes and values of literature and the "better" or "worse" readings that a teacher customarily finds in student work (despite grade inflation, not every paper deserves an A).[20] Indeed, the most interesting moments in a classroom occur not when a student is obviously wrong about some textual detail (as does happen) but when students find themselves having disagreements that cannot be resolved by appeals to "the text itself."

The fact that interpreters may have interesting, productive disputes about textual meaning does not prevent agreement among practitioners whose authority has been accredited in various ways that some interpretations are less credible than others. The line between what counts as a right or wrong reading is not carved in stone and may change as new methods of interpretation emerge (and old methods are retired), but the fact of interpretive disagreement does not mean that all interpretations are equally "correct." The aim of neuroaesthetic inquiry should be to explain and explore the neurobiological implications of such conflict (why and how the brain is open to opposing interpretations as it negotiates the competing demands of constancy and flexibility), not to end it. Here and elsewhere, conversations between neuroscience and the humanities can and should respect their disciplinary differences.

The most important potential gain for the humanities from conversations with neuroscience is, I think, a refocusing of attention on several core questions of common interest: What happens when we read literary (and nonliterary) texts? What are the characteristics of aesthetic experiences? How is the interpretation of literature related to other epistemological processes, and what are we to make of the conflict of interpretations? These are issues of concern to neuroscience, and they are also basic questions for the humanities.

They are questions that have by and large been sidelined, however, by the dominance over the last few decades of various contextualist approaches to criticism (political, social, historical, and cultural). Unfortunately, the sidelining of these basic questions has gone hand in hand with a marginalization of the lettered humanities in the academy, as practitioners from other disciplines have found less and less of interest in what we are doing even as, ironically, literary critics have felt themselves to be widening their scope of inquiry. The reasons for these developments are complicated and have, perhaps, as much to do with an overestimation in the social sciences of the value of quantitative methods as with the humanities' abandonment of literary and aesthetic questions. Changes have begun to occur in the humanities, and there is evidence on many fronts of a return of interest in reading, aesthetics, and form.[21] Taking up conversations with the neuroscientific community about matters of mutual interest may help the humanities rediscover the core disciplinary concerns with reading, interpretation, and aesthetics, about which we can speak with authority. What the humanities have most to gain from this interdisciplinary work, then, is a recovery of our disciplinary identity.

The divide between the so-called two cultures will not be bridged by creating a third culture or even a fourth culture (as some have proposed) to mediate their differences.[22] Rather, this gap can be crossed by conversations between parties on both sides who share common interests and find mutual benefit from discussing them. A good place to start this conversation would be the question that guides this book: How do we read, and how does literature play with the brain? The experience of reading is a place where the humanities and neuroscience can productively meet—with much to be gained on both sides—without betraying their core values or identity, but by speaking to each other from positions of disciplinary strength and integrity. For the humanities, this would mean going back to basics, not in the sense of returning to a traditional canon but by taking up central questions that have a long history and that have been (and remain) crucial to our enterprise. Interdisciplinary research works best when parties with different methods and knowledge find reason to collaborate, share, and exchange because the other's perspective and expertise offer something they need. Refocusing on questions of common interest with neuroscience may help the humanities get out of their isolation and end their widely perceived irrelevance.

Unfortunately, given the explosion of interest in both worlds about what has come to be called *neuroaesthetics*, there is less conversation than one might expect between neuroscientists and literary theorists about matters of mutual interest.[23] Numerous reasons can be given for this, beginning no doubt with the suspicions about the "other" on both sides of the divide between the two cultures (among the scientists, about the seeming lack of methodological constraints in the humanities, and among the humanists, about the unapologetic reductionism of science). Organizational barriers also stand in the way of collaboration. Despite the popularity of interdisciplinary research, the silos in which universities are structured discourage scientists and humanists from interacting. There are also genuine, nontrivial disciplinary obstacles to mounting and sustaining meaningful dialogue. It is hard, after all, to master enough of the language of either neuroscience or literary theory to engage in meaningful acts of translation.

These obstacles are all the more reason to be clear about the common ground on which such conversations might take place. Although not the only candidate, the experience of reading literary works is an obvious area of shared interest. How readers read and how literary works take up and manipulate these processes in various ways are matters about which neuroscientists and literary critics can say useful things to one another, even if their conversations will not solve the hard problem and explain how neurological functions produce conscious experiences.

When responsibly informed by aesthetic theory, the experiments of neuroscience may help to clarify literary experiences that theorists since Aristotle have puzzled over, such as how "pity" and "terror" combine in the viewer of a tragedy in order to induce "catharsis." The neurological correlates of this experience may not fully explain its mystery, but (as I try to show in chapters 4 and 5) neurobiological studies of how the brain simulates body states can clarify how these and other aesthetic emotions occur. Whether catharsis is purgative is also a matter on which neuroscientific studies of the effects of violent representations can shed light. Similarly (and as the next chapter shows), neuroscientific investigations of how we learn to read can help to explain why play has so often been associated with aesthetic experiences and why play both pleases and instructs. The extensive scientific work on the neurobiology of vision also helps to clarify and explain (as I demonstrate in chapter 3) the long-discussed paradox of interpretation that it is circularly dependent on our expectations. In these and other areas that this book analyzes, parallels between neuroscience and literary theory can illuminate processes and experiences that have intrigued both worlds. These comparisons do not reduce either discipline to the other; literary theory does not become a subfield of neuroscience, and vice versa. The wager, rather, is that problems of interest to each can be clarified with the insights and methods distinctive to the disciplinary perspective of the other. Talk across the explanatory gap may benefit both fields.

One of the biggest fallacies threatening neurobiological work in aesthetics is the monistic assumption that "*the*" aesthetic experience exists in a singular, unique form that (allowing for variations between visual, aural, and linguistic arts) can be definitively correlated to specific regions of the brain and identifiable neurological processes. The issue here in neurobiological terms is the extent to which brain functions can be localized. Across the explanatory gap, in aesthetic terms the question is whether distinctive, univocal markers can be identified to demarcate literary from nonliterary phenomena or aesthetic from nonaesthetic experiences.

Neuroscientists exploring art and literature generally agree that there is no "art neuron" and that aesthetic experience is more complex and distributed more widely over the brain than, say, color perception or facial recognition, which have specific locations in the cortex.[24] Although the notion of an art neuron is absurd, it might seem a tempting hypothesis because individual neurons can take on particular functions. For example, in a case that has become understandably notorious in the neuroscience literature, R. Q. Quiroga discovered a neuron in the anterior temporal region of an epilepsy patient that fired solely in response to the Hollywood film and television star Jennifer Aniston, whether the patient was shown a photograph or a line drawing or even her written name.[25] This neuron was so particular that it did not respond to images of Aniston with Brad Pitt. The "Jennifer Aniston neuron" shows, however, that brain functions are both anatomically localizable and open to experiential variation, and this combination of localization and variability is even more complex with phenomena (like the experience of art and literature) that integrate widely dispersed cortical regions and even (as with emotional, visceral responses like pity and fear) cross the brain-body divide. This complication is reflected in the current consensus, which neuroscientist David Keller succinctly summarizes as follows: "The neural activity which enables human beings to appreciate artworks does not take place in a circumscribed brain region, but is instead temporally and spatially distributed."[26] Given how the brain processes information through reciprocally connected networks, this is not surprising and is not unique to art.

Curiously, after granting that strict localization is not the goal, one finds that the same disagreement that occurs again and again in the history of criticism also divides researchers in neuroaesthetics as to what they should be looking for. On the one hand, the eminent French neuroscientist Jean-Pierre Changeux argues that "harmony, or *consensus partium*," is the mark of the aesthetic, manifested by "formal appropriateness, where the unity of the whole triumphed over the multiplicity of its parts." As his colleague Stanislas Dehaene explains, "A work of art becomes a masterpiece when it stimulates multiple distributed cerebral processes in a novel, synchronous, and harmonious way."[27] According to this view, unity, symmetry, and harmony are the signs of art and the hallmarks of aesthetic experience. On the other hand, the valuable insight that the neuroscientist Ramachandran came up with in his back-of-the-envelope musings about fundamental neuroaesthetic principles is that art's distortions often matter more than its pleasing syntheses: "The purpose of art . . . is not merely to depict or represent reality . . . but to enhance, transcend, or indeed even to *distort*" it through techniques that may be disruptive, disjunctive, and nonharmonious and in this very way "powerfully activate the same neural mechanisms that would be activated by the original object."[28]

This conflict over whether the aesthetic experience is characterized by harmony, unity, and synthesis or distortion, disruption, and dissonance can be found repeatedly in the history of aesthetics. It is evident most notably, perhaps, in the dispute between classical conceptions of art as balanced, rule governed, and symmetrical and Romantic valorizations of rule-breaking, original genius and idiosyncratic particularity.[29] This opposition recurs again and again, and it calls into question the notion that there is a uniquely aesthetic emotion or that aesthetic experience is a singular, uniform, and identifiable phenomenon. A short survey of the conflict between aesthetics of harmony and aesthetics of dissonance would include the Prague structuralist Roman Jakobson's identification of the poetic function with the predictable regularities of rhythm and rhyme, as opposed to the Russian Formalist Viktor Shklovsky's definition of art as "defamiliarization," the "making strange" of habitualized perception; the New Critics' emphasis on "organic unity" achieved by synthesizing contradictions through "paradox" versus the deconstructionist claim that literariness is marked by a refusal to resolve irreducible differences; or the conflict in Marxist aesthetics between Lukács's preference for totalizing representations that make coherent sense of the particularities of the real and Brecht's disruptive, distancing techniques of "alienation" (*Verfremdung*). This list could go on and on. The conflict between harmony and dissonance in aesthetics is fundamental and unsurpassable.

The polarity between harmony and dissonance is, it would seem, a universal feature of the aesthetic. If that is so, it is a universal that reflects and helps to explain the fundamental heterogeneity of art. The paradox of art is that it is both universal and relative, a singular feature of human culture and a multiple, variable phenomenon that is not reducible to a uniform, unifiable set of properties. One reason for this paradox is that the brain seems to be hard-wired to respond to harmonious and dissonant forms, but what counts as harmony or dissonance is contingent and historically variable. Not surprisingly, this paradox has been studied extensively by neuroscientists of music, and I discuss in chapter 2 what their experiments show about the universality and cultural relativity of responses to musical *consonance*. Although I use the terms metaphorically to refer to effects of synthesis and disruption in all art forms, a large and growing body of neuroscientific research on the cortical bases of musical harmony and dissonance suggests that they are both biologically universal and culturally relative because of the constraints and plasticity of the auditory system.

In literature and other arts, the aesthetics of harmony take on a variety of manifestations, all emphasizing balance, symmetry, and unity, but in different ways that reflect the assumptions, beliefs, and values of the community in question, so that what comes to be valued as harmonious form can vary historically and culturally. Conversely, the power and purpose of various aesthetics of dissonance typically have to do with how their artistic programs attack, question, and subvert prevailing notions of artistic harmony, and so dissonant art is similarly diverse and historically contingent (even if it is universally intent on disruption and transgression). Neurobiologically, aesthetics of harmony appeal to and reinforce the brain's need for synthesis and pattern, whereas dissonance serves the purpose of keeping the brain flexible and open to change, combating the rigidities of habit. These are universal functions and needs that literature and art play with in an ever-varying array of forms.

Harmony and dissonance are interdependent aesthetic values. This can be seen in the oft-observed fact that one generation's dissonant protest frequently becomes a later generation's target as what was originally perceived as disruptive and transgressive becomes assimilated and conventional. Flaubert's novel *Madame Bovary* was prosecuted as an outrage to bourgeois morality when it first appeared in 1857, for example, but came to be valued in the modern period as an aesthetic ideal through which the quotidian mediocrity of everyday life was transcended by its transformation into beautiful linguistic form. In yet a further twist, after the New Critics praised Flaubert's irony for its synthesizing beauty, a later generation of deconstructive readers argued that this same strategy subverted any endorsement of a particular set of values, and "irony," which was once seen as creating order out of chaos, came to be viewed as a means of questioning everything normative. When Cleanth Brooks claims that "the language of poetry is the language of paradox," it is because of the ability of irony to forge unity out of multiplicity, but Paul de Man values irony (must we mean what we say?) because it disrupts and destabilizes boundaries that seem to make meaning determinate.[30] Many more examples could be given of how aesthetics of harmony and aesthetics of dissonance are mutually defining and can take on different values and even switch places over the history of art and criticism.

Harmony and dissonance in art are defined not only by each other but also by their opposition to a third concept—noise—and this relationship is also historically and culturally changeable. Like harmony and dissonance, noise has characteristics that are both universal and relative. The human brain's capacities for synthesis are not unlimited, just as there are spectra of light and frequencies of sound that escape its ken (although other animals can process them). Some stimuli simply are not noticed. If they are perceived but cannot be processed, they are noise.

Noise cannot be assimilated by the brain, either because it is too disjunctive to be synthesized (the jarring chaos of random sound) or because it is too uniform to be differentiated meaningfully (so-called white noise, which can function as a soothing background). Noise is a neurobiological constant because it reflects the limits of the organism's sensory apparatus, but it is also for this reason a variable that can change not only from one species to another but also within a species and even within individuals as the organism's capacities and circumstances vary. (The classical music I enjoy, for example, has been used to keep adolescents from loitering at convenience stores, although they may later learn to love Beethoven, even as the musical tastes of my own children have restructured the noise boundaries in my brain.) The triad harmony-dissonance-noise is a universal structure whose points can be occupied by different values and forms. Indeed, the fact that it is a triad of reciprocally constituted categories explains why and how it is open to historical, cultural variation.

It is also important to remember that the opposition between harmony and dissonance is not unique to art, and this in turn points out the error of assuming an absolute discontinuity between aesthetic and nonaesthetic experience. As the neuroaestheticians Martin Skov and Oshin Vartanian observe, neurobiological processes linked to aesthetic phenomena "are common rather than unique, and are obviously evoked when we create paintings and watch movies, but also when we embrace a loved one."[31] Experiences of harmony and dissonance occur not only when we read great poetry but also as we go about our daily lives. Attempts to identify the distinguishing features of art typically run aground not only because the aesthetic is so heterogeneous but also because the proposed markers turn out to be evident in other, ordinary phenomena. For example, in a recent influential attempt to identify "literary universals," Patrick Colm Hogan proposes that literature differs from "ordinary language" because it "maximizes" linguistic patterns and thereby foregrounds the features of an expression.[32] This recalls the definition of the *poetic function* by Roman Jakobson (on whom Hogan draws), through which a message calls attention to itself by parallelisms and repetitive patterns. But Jakobson notes that foregrounding of this kind is pervasive in ordinary language, as in the political slogan "I like Ike" or in advertising jingles, as in the rhyme admired by that inveterate adman Leopold Bloom in *Ulysses*: "What is home without / Plumtree's Potted Meat? / Incomplete."[33] Foregrounding alone is not a necessary or sufficient marker of the literary.

The boundary between the literary and the nonliterary is inherently fuzzy and permeable, and this vitiates the proposal of the otherwise perceptive and extremely knowledgeable neurocritic David Miall that "the resolution with which scanning occurs should enable us to discriminate the structures that distinguish a literary from a nonliterary response."[34] Aside from the questions this claim raises about the precision of brain-imaging technologies, it assumes a more clear-cut distinction between aesthetic and nonaesthetic experience than can be theoretically justified. The experience of great art may seem so special and unique that one assumes it must be a distinct, specifiable phenomenon (an assumption many philosophers and critics have made over the long history of aesthetics), but art can give rise to an extraordinarily diverse array of experiences, and it would actually be surprising if a phenomenon as complex and historically various as art were reducible to a unified, singular set of conditions.

It is sometimes thought that art announces its presence by provoking an aesthetic emotion in the perceiver. The most influential modern articulation of this view was provided by Virginia Woolf's brother-in-law Clive Bell, who claimed that "there is a peculiar emotion provoked by works of art" in response to what he famously called "significant form," an aesthetic emotion that "transports us from the world of man's activity to a world of aesthetic exaltation. For a moment we are shut off from human interests; our anticipations and memories are arrested; we are lifted above the stream of life."[35] Bell's description of the aesthetic experience is, of course, a legacy of Kant's theory of beauty, which sets the aesthetic experience apart from ordinary perception and judgment or from pragmatic pursuits in an ideal sphere of "interested disinterestedness."[36] The emotions produced by aesthetic experiences are more wide-ranging and various than this rather sterile formulation suggests, however. There is no single aesthetic emotion any more than there is a single aesthetic experience or a single definition of art. But aesthetic emotions do differ from the everyday emotions they reenact, because they have an "as if" dimension. The pity and fear we feel in response to a tragic hero's fate both are and are not the compassion and terror we might experience in a similar real-life situation. The pretense of the "as if" does distinguish reenacted emotion from the real thing, but the "as if" of aesthetic emotions also draws on lived, embodied experience and re-creates it. One reason for the heterogeneity of art and the difficulty of drawing clean boundaries between art and non-art is that the "as if" of aesthetically reenacted emotions can take so many forms (and "as if" experiences also abound in everyday life, whenever we have a vicarious thrill or identify with another's joy or suffering).

Art and aesthetic experience are as diverse and wide-ranging as human emotional and cognitive life. Rather than seeking the neural correlates of the supposedly distinctive if elusive aesthetic experience, a better approach is to acknowledge the diversity of the aesthetic and then to explore how its variety is linked to the full range of the brain's functions and locations. This kind of mapping of the particular relations between aesthetic experiences and neurological processes and regions is perhaps less intellectually exciting than a unified "theory of everything" would be (at least to the monists among us, who delight in grand syntheses). But it is a more accurate reflection of the complexity and variety not only of art but also of the brain (and exploring these multiplicities has many pleasures for the pluralistically minded like myself). What this mapping may find—my argument, indeed, is that it will—is that specific theories about what constitutes the aesthetic experience are linked in identifiable ways to different, particular neurological processes and the cortical regions that these interactions connect. How various views about aesthetic experience key in on particular brain functions is one of the recurrent themes of this book.

The goal of neuroaesthetics should not be the discovery of a singular neurological correlate for art, but a differentiated understanding of the neurobiological underpinnings of various aesthetic experiences (each defined as aesthetic in a particular way). These neurological processes will show that an array of interesting similarities connect these aesthetic experiences and the ordinary emotional, cognitive, and perceptual experiences that are being mapped with ever-increasing subtlety and complexity by contemporary imaging technologies. The result to aim for is a better understanding, not of the absolute uniqueness of art, but of the complex continuities and discontinuities between aesthetic experiences and the everyday life of our species. To see how art and aesthetic experience are connected to the workings of the brain is thus to understand how deeply rooted they are in the life of our species.

These links between art and the brain have so far been relatively unexplored by literary theory. The so-called cognitive revolution in literary studies has taken primarily two forms, neither directly connected to contemporary neurobiology.[37] First, reflecting the recent dominance of historical, social, and cultural approaches in literature departments, important comparative studies have charted relations between various past literary movements or genres and the prevailing cognitive science of the time.[38] Second, a growing number of literary studies based on psychologically oriented cognitive science have applied experimentally based theories about such matters as textual comprehension, empathy, and the reading of other minds.[39] This reliance on cognitive science, which explores how the "mind" knows the world, rather than neurobiology, which focuses on the structure and functioning of the brain, is no doubt understandable, since it is easier to move to literary phenomena from psychological theories about mental processes than it is to bridge the gap between neurological mechanisms and lived experience.

How (or even whether) the "mind" and the "brain" are related is, however, an extremely controversial question, and some hard-core materialists in the neuroscience world are deeply skeptical that the "mind" is anything but an epiphenomenon. It would seem reasonable to think of these as two approaches to common problems that may eventually converge. As the neuroscientist Stanislas Dehaene puts it, "A direct one-to-one relation exists between each of our thoughts and the discharge patterns of given groups of neurons in our brain—states of mind *are* states of brain matter."[40] As the neurophilosophers Andrew Brook and Pete Mandik explain, however, it is possible to interpret this assertion in two very different ways: either as crediting the autonomy of both mind and brain as parallel and distinctive methodological concepts, or as regarding mind as a temporary, provisional, and ultimately disposable construct along the way to the ultimate reduction of mental states to their physical basis. There is also, they note, a third position, which they call radical *eliminativism*. According to this view, "psychological theories are so riddled with error and psychological concepts are so weak when it comes to building a science out of them (for example, phenomena identified using psychological concepts are difficult if not impossible to quantify precisely) that psychological states are best regarded as talking about nothing that really exists."[41]

Unfortunately, given this controversy, the reliance of contemporary literary criticism on cognitive science may be an obstacle to collaborations with neuroscience, even if, as I show throughout this book, there are a number of important convergences between these two fields. For example, as the next chapter shows, the neuroscience of reading draws important evidence from cognitive psychology's studies of word processing and language comprehension. Similarly, as I explain in my exploration of the brain's social skills in chapter 5, contemporary mirror-neuron research relies heavily on psychological observations of infant imitation to explicate the mechanisms of intersubjectivity. Nevertheless, despite these and other convergences, productive dialogue between literary studies and neuroscience may be facilitated if the mind-brain dispute can be bypassed and a way to correlate aesthetic experiences and neurological mechanisms that does not rely on assumptions about the mind can be found.

A path forward might be provided by *neurophenomenology*, a research program initiated by the Chilean neuroscientist Francisco J. Varela to study the "mutual constraints" through which "lived experience and its natural biological basis are linked."[42] As a summary by his followers explains, "Neurophenomenology aims not to close the explanatory gap (in the sense of conceptual or ontological reduction) but rather to bridge the gap by establishing dynamic reciprocal constraints between subjective experience and neuro-biology."[43] The founder of modern phenomenology, Edmund Husserl, and other philosophers in his tradition sought to provide rigorous descriptions of the structures of lived experience by *bracketing*, or putting out of play, the assumptions of the *natural attitude* about subjectivity and the object-world. Using these methods and applying findings to which they have led, neuro-phenomenologists seek to correlate neuroscientific results from experiments about brain functioning with sophisticated, precise, and reliable first-person accounts of consciousness.[44]

For example, Varela argues that Husserl's complex, nuanced account of the *retentional* and *protential* horizons of our lived experience of the ever-passing present is supported by and in turn helps to explain the emerging consensus in experimental neuroscience that brain regions are interconnected in a re-ciprocal fashion that cannot be accounted for by the old sequential arrange-ment of the computer metaphor.[45] As I explain in detail in chapter 4, these multidirectional relations are characterized at the neuronal level by patterns of activation, relaxation, and oscillation that require a conception of tem-porality more like Husserl's description of the ever-emerging and receding "horizonal present" than a linear sequence of pointlike moments. The recip-rocal determination and mutual interaction of neuronal assemblies across sometimes widely dispersed regions of the brain are coordinated through temporal patterns of oscillation that have durational width. The phenomeno-logical paradoxes of lived time—that the past and the future are both present and absent across the horizons of the passing moment—have a neurobiologi-cal foundation, then, and the peculiarities of a nonsequential, reciprocally determined model of brain interactivity are not experientially counterintui-tive. The temporal structure of reciprocal neuronal interactions on the one hand and the lived experience of retentional and protential (past and future) horizons on the other are not reducible to each other. There is still a hard problem of how one gives rise to the other across the explanatory gap divid-ing the temporality of brain processes and our consciousness of time, but the explanations of how time works at these different levels help to clarify the paradoxes and complexities evident at both.

At least to date, neurophenomenology has contributed little to neuroaes-thetics. This is especially unfortunate because of the rich tradition of phe-nomenological research on aesthetics, interpretation, and reading to which Husserl's work gave rise, from the early analyses of his student Roman Ingarden of the literary work of art as an intentional, intersubjective structure and its "concretization" in reading through the hermeneutic investigations of Heidegger, Gadamer, and Ricoeur into the structure of understanding and interpretation (following the lead of Husserl's theory of intentionality), to the reception theory of the Konstanz Poetics and Hermeneutics group led by Wolfgang Iser and Hans Robert Jauss.[46] For example, in ways that I try in chapter 4 to make clear in some detail, the correlations between neural and phenomenological accounts of the temporality of cognition as nonlinear, horizonal, and reciprocal have important parallels not only in Heidegger's description of the hermeneutic circle as a structure of anticipation but also in Iser's conception of reading as a to-and-fro process of consistency building. These parallels suggest ways in which the lived experience of reading and its aesthetic manifestations relate to fundamental neurological processes that research on the brain has identified.

For example, as I show in chapter 3, the hermeneutic circle—the paradox whereby an understanding of the parts of a text depends on an anticipatory sense of the whole to which they belong—turns out to have deep foundations in the cognitive functioning of the brain. Similarly, as I explain there and also in chapter 2, there is extensive neurological evidence about how the brain interprets shapes and words that is consistent with the phenomenological view of reading as a process of filling in textual indeterminacies and building consistent patterns, a process open to opposing results (so that readers may disagree about what a text means). Neurological research on the brain's response to ambiguous figures and the possibility of multiple interpretations is consonant with phenomenological theories of multiple meaning and conflicting readings.[47]

These parallels should be of interest to literary criticism for a variety of reasons, not least because they suggest that the experience of reading is not the epiphenomenon that some contextually oriented social and political critics make it out to be. Describing themselves as materialists, a sizable population of skeptical cultural critics tend to view consciousness and subjectivity with suspicion, and for them, phenomenological theories of interpretation and aesthetic experience are best misguided and at worst illusory.[48] Clinical and experimental evidence amply and conclusively shows, however, that there are material neurobiological bases for phenomenology's aesthetic theories. Reading and aesthetic experience are becoming of interest again to at least some in the humanities, and the biological reasons for taking these phenomena seriously provide materialist arguments to counter the skepticism of the materialists. On the other side of the explanatory gap, these connections should also be of interest to neurobiology because they suggest ways in which sometimes rather rudimentary neurological evidence might be relevant to phenomena (like reading and the aesthetic experience) that are as yet much too complex and subtle for currently available experimental technologies to map. Phenomenological descriptions of these processes can suggest questions deserving of experimental neuroscientific inquiry to fill in some of these areas of obscurity.

Interestingly, phenomenological descriptions of reading have produced theories in line with both the aesthetics of harmony and the aesthetics of dissonance. This is a useful contradiction, not a sign of confusion, and it is not at all surprising. After all, given that generations of readers have reported aesthetic experiences of both kinds, phenomenological accounts of the lived reception of art would need to explain how such a fundamental, recurring divergence can occur and not simply dismiss as baseless or wrong a whole history of experiences on one or the other side of this divide. Such an explanation is suggested by the disagreement between the pioneering phenomenological aesthetician Roman Ingarden and the later Konstanz School theorists of reception (who draw on his work) about whether aesthetic experience is constituted by a harmonization of felt values or a disruption of the reader's expectations. Ingarden identifies four interrelated "strata" that are held ready for the reader of a literary work and are "concretized" in its cognition: "linguistic sound formations" (phonetic patterns), "meaning units" (at the word and sentence level), "represented objects" (the people, places, and things constituted by the semantic level), and the work's "schematized aspects" (the perspectives through which these states of affairs are displayed, with much necessarily left "indeterminate" or hidden, for the reader to fill in). According to Ingarden, the reader achieves an aesthetic experience by forging a "polyphonic harmony" of the "value qualities" held ready by each stratum.[49]

While building on Ingarden's investigations, especially his finding that literary works are full of indeterminacies (*Unbestimmtheitstellen*) that readers may fill in variously (or not) according to their presuppositions, interests, and past experiences, Wolfgang Iser takes issue with "his adherence to the classical idea of art" and instead finds value in the reader's experiences of disruption, disjunction, and discontinuity: "Literary texts are full of unexpected twists and turns, and frustration of expectations. . . . Indeed, it is only through inevitable omissions that a story gains its dynamism" and allows us "to bring into play our own faculty for establishing connections," thereby leaving open the possibility that readers will construct a text's world in different, even radically opposing ways.[50] Iser's colleague and fellow founder of the Poetics and Hermeneutics group, Hans Robert Jauss, takes this emphasis on the value of frustrated expectations one step further and proposes to measure "the artistic character of a literary work" by the degree to which it upsets, disappoints, and otherwise challenges the reader's familiar conceptions of art, morality, and other matters. According to Jauss, what he calls "'culinary' or entertainment art [*Unterhaltungskunst*] . . . can be characterized . . . as not demanding any horizonal change, but rather as precisely fulfilling the expectations prescribed by a ruling standard of taste" or morality.[51] By contrast, great works of art (on this view) challenge familiar conceptions and as a result often are met with outrage, misunderstanding, or critical condemnation, but then, paradoxically if also understandably, they risk losing their effect to the extent that they become familiar and normative. It often happens that canonical works must be defamiliarized or read against the grain, Jauss argues, to regain their edge.

Rather than taking sides in the conflict between the aesthetics of harmony and the aesthetics of dissonance, neuroaesthetics should ask how these accounts of the pleasures and challenges of having expectations met or thwarted are related to the brain's processes of comprehension. Neurological functions of special importance to understanding harmony and dissonance would be those associated with pattern recognition and the temporality of the multidirectional signaling across cortical regions through which networks of neurons are joined. Other important topics would be the neurological correlative to surprise, the brain's processing of unfamiliar as opposed to familiar experiences, and the neural equivalents of integration and discontinuity. These are some of the places where the potential for art resides in the brain. The first step in exploring them, at least in their literary manifestations, is to ask how the brain learns to read.

A cautionary note about language and terminology is in order before we begin. The locution "the brain learns" is a metaphor; as such, it is both illuminating and distorting because it misstates the literal facts of the matter. As I point out repeatedly in what follows, the brain is not a homunculus—there is no "little man in the machine" inside our heads—and it is consequently misleading to attribute goals, purposes, and directions to this biological organ. Once one is alerted to this problem, however, one sees these locutions everywhere in the literature of neuroscience. For example, "the brain abhors internal anomalies . . . and will often go to absurd lengths to explain them away," or the brain is "a mimic of the irrepressible variety" that "builds maps" and "relates them," or "the only work that the brain must do is whatever it takes to assuage epistemic hunger—to satisfy 'curiosity' in all its forms."[52] Even the neuroscientist-philosopher Alva Noë, who rightly warns, "You are not your brain," cannot help but talk about this anonymous organ in personified terms, as if it were an agent pursuing various purposes and intentions: "The brain's job is that of facilitating a dynamic pattern of interaction among brain, body, and world."[53]

Part of the problem here is linguistic. Metaphors reveal but also disguise and mislead. It is impossible—and stylistically unwise—to write without them, and that is why some deconstructively minded critics put slashes through expressions to indicate that they are being used "under erasure," in full cognizance of their necessary distortions.[54] Putting subjects and verbs together tends to make the noun seem like an agent engaged in purposive action, even when an organ like the heart or the kidneys is involved (for example, giving these organs moral and social agency, Wikipedia says that "the heart is responsible for pumping blood" and "kidneys serve essential regulatory roles").

But there is also an important philosophical issue here, as Paul Ricoeur noted in his illuminating extended conversation with Jean-Pierre Changeux about human nature, ethics, and the brain. According to Ricoeur, the "discourses" of morality and neuroscience "represent heterogeneous perspectives" that "cannot be reduced to each other or derived from each other. In the one case it is a question of neurons and their connection in a system; in the other one speaks of knowledge, action, feeling—acts or states characterized by intentions, motivations, and values. I shall therefore combat the sort of semantic amalgamation that one finds summarized in the oxymoronic formula 'The brain thinks.' " To which Changeux quickly and defensively replied: "I avoid using such formulas."[55] Such linguistic purity is not always possible or desirable, however. Metaphors do serve useful rhetorical purposes.[56] But how agency arises out of electrochemical, neurophysiological processes in the brain is one of the essential mysteries that make the hard problem difficult and perhaps unsolvable. The brain is a complex, fascinating biological organ, not an agent with goals and purposes. But some of its capabilities are so amazing that one thinks it must be animated with intentions. It isn't. There is no person inside your head, and you are not your brain. But the brain is capable of astonishing transformations. And foremost among them are the processes by which (if you will pardon the metaphor) it learns how to read.

CHAPTER 2

How the Brain Learns to Read and the Play of Harmony and Dissonance

Aesthetic experiences in different arts share some features and qualities, although certain experiences are of course specific to a particular art, as are their neurological and physiological correlates. Harmony and dissonance can characterize music, visual art, and literature, for example, but there will necessarily be differences in the neurological processes and cortical areas associated with these experiences, as there are differences between the systems connected with hearing, vision, and reading. These relatively obvious points are worth making before we launch into an exploration of the neuroscience of reading, because of the particular complications and peculiarities of this hybrid phenomenon. Unlike vision and hearing, reading is not a "natural" function with its own intrinsic, dedicated neurobiological systems activated at birth. The aesthetics of reading share much with the visual arts and music, in part because the graphic and phonetic processes of word recognition depend on vision and hearing. The appreciation of the visual arts and music no doubt also requires learning—the acquisition of skills and conventions through which one "reads" a painting or interprets a symphony—and vision and hearing themselves are as much historical as they are natural because they are the product of long processes of evolution. But more than either visual perception or hearing, the comprehension of written linguistic texts is a neurological hybrid that draws on an array of brain processes that evolved in the first place for other purposes and are primarily dedicated to other functions.

How the brain learns to read is a complicated, in some ways paradoxical process that reveals much about a number of important issues in neuroscience. For example: What is the relation between the brain's fixed, inherited characteristics and its plasticity, its capacity to change, adapt, and develop? To what extent are certain cognitive functions localized in particular cortical areas, and to what extent is brain functioning distributed over different regions that reciprocally interact and are malleably configured? How is the brain organized? Is it like a computer with a central controller, or a parallel-processing network, or a chaotic, unstable array of ever-changing interactions? Explaining the neuroscience of reading may begin to answer some of these questions and thereby serve as an introduction to the workings of the brain for readers to whom this is unfamiliar terrain. This in turn will set the stage for exploring the neurological bases of some important aesthetic phenomena (like the play of harmony and dissonance) that can be found not only in literature but also in other arts and that in many instances are live issues in contemporary neuroscience.

The central neuroscientific fact about reading is that it is a relatively late development in the history of our species that could only emerge by exploiting preexisting neurological systems. The brain is structurally predisposed to acquire language, and everyone learns how to speak in early childhood unless physiological impairments prevent this natural development.[1] Reading does not come naturally, however, and not everyone learns how to do it. Estimates vary about when to date the birth of reading and writing, but it is likely to have happened no more than six millennia ago. The historian Steven Roger Fischer plausibly argues that "reading in its true form emerged when one started to interpret a sign for its sound value alone within a standardized system of limited signs," and he claims that the "sign became sound—freed from its system-external referent [as in pictographic representation]—in Mesopotamia between 6,000 and 5,700 years ago."[2] This deciphering of coded, conventional marks linking sound and sense is consistent with how modern neuroscience understands reading. In order to construe the meaning of written signs, the reading brain must translate phonetically and graphically linked units (*phonemes* and *graphemes*) into formal structures (*morphemes*) with semantic significance. When the brain learns to perform this act of translation, reading is born.

A few thousand years is trivial in the history of evolution and not nearly long enough for genetic transformations of the brain that would be necessary for this capacity to develop through Darwinian *natural selection*. What had to occur is what the neuroscientist of reading Stanislas Dehaene memorably calls "neuronal recycling," the repurposing of "a cortical territory initially devoted to a different function."[3] Every new human being must learn to read by adapting genetically inherited circuitry to uses for which it did not originally evolve, and some of the difficulties encountered by beginning readers, as well as some of the differences in how easy this learning is for readers in different languages, are traceable to mismatches between the requirements of decoding written signs and the neurological systems that must be converted to this unnatural act.

Clinical and experimental evidence suggests that this conversion occurs in a region of the brain devoted to the recognition of visual forms. The first indication of a *visual word form area*, or VWFA, dedicated to reading came in the late nineteenth century when a patient who suffered a minor stroke lost the ability to read, while retaining the capacity to speak and recognize objects other than written words. Modern brain-imaging technology has located an area of the lower left hemisphere that is activated in response to written signs (but not to spoken words, which trigger a different area [see fig. 2.1]). The neuroscientist whose laboratory has done the most prominent work on the VWFA, Dehaene calls this area "the brain's letterbox" (53) and reports that it can be found in the rear visual cortex, on the underside of the brain, sandwiched between the region devoted to recognizing objects and the neurons keyed to faces. This finding is somewhat controversial because the VWFA is not homogeneous and still bears traces of other activity (as one would expect because it is repurposed from other functions), but evidence of its existence and its role in word recognition is compelling.[4]

Brain-imaging experiments show that the VWFA is activated by all alphabetic systems, by Chinese as well as Roman characters and by both the kanji and kana scripts used by Japanese.[5] These experiments reveal the niche in the architecture of the brain that has been redirected to a specific cultural purpose that arose too quickly for biological evolution to accommodate with genetic changes, and this is powerful evidence for the mutual accommodation of nature and culture assumed by the hypothesis of neuronal recycling. The universality of the niche across cultures with different alphabets is evidence of the restrictions of pre-given cortical structures, even as the conversion of a particular region of visual-recognition neurons to an unnatural, learned, culturally variable activity shows the plasticity and adaptability of the brain.

The selection of this area of the brain for recycling is not accidental but seems to have been a consequence of its role in invariant visual object recognition. The ability to identify the same object, place, or person under different conditions—changes in lighting, distance, orientation, and so forth—is absolutely necessary for human survival. The ability to recognize a visual form invariantly under changing conditions is crucial not only for perceiving ob-

"Pigeon" Read or Spoken

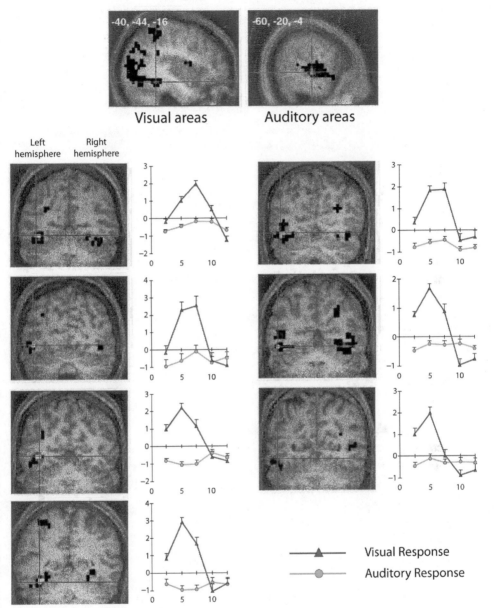

Visual areas   Auditory areas

Left hemisphere   Right hemisphere

Visual Response

Auditory Response

Figure 2.1. Activation of the visual word form area (VWFA) in fMRI images of seven adult readers. Spoken words did not activate this area. Adapted by permission from Stanislas Dehaene et al., "The Visual Word Form Area: A Prelexical Representation of Visual Words in the Fusiform Gyrus," *NeuroReport* 13.3 (4 March 2002): 321–25.

jects in the external world but also for identifying words rendered in different shapes: written in upper or lower case, in different fonts, and even in cursive script (within limits, of course, that my own penmanship constantly tests). It is a defining feature of reading that we are able to recognize the word *radio* even when it is written as *RADIO* or even *RaDiO*, and this ability is an example of the invariant visual perception of objects.[6] The capacity of neurons in the letterbox area to ignore variations in case and recognize the self-identical word written in different shapes is evidence that cortical functions that originally evolved for invariant visual object recognition have been adapted to the artificial, culturally specific purposes of reading conventional graphic signs.

Somewhat speculative but highly suggestive evidence about recurrent preferred shapes across different alphabetic systems further demonstrates the deep neurological link between reading and invariant object recognition. Some of these constraints are attributable to the physiology of vision and the capacity of the retina to register shapes. Others, however, seem to suggest that writing systems developed by drawing on visual markers of object invariance to which the brain had become accustomed before the advent of reading.

This, in any case, is the hypothesis put forward by the evolutionary neurobiologist Mark Changizi based on a comprehensive comparative analysis of alphabetic forms. According to his research, visual signs in writing systems "have been culturally selected to match the kinds of conglomeration of contours found in natural scenes because that is what we have evolved to be good at visually processing."[7] Through a statistical analysis of images of naturally occurring shapes, Changizi found that objects often form T- or L-like patterns when they are viewed together, whether placed next to each other or one in front of the other, partially blocking one's vision of the one behind. Sometimes, but more rarely, an X shape occurs, as when one branch of a tree crosses another. Three lines forming a triangle seldom occur in a natural scene. According to Changizi, the frequency with which these shapes occur in nature is, surprisingly and strikingly, closely comparable to the distribution of similar shapes in the world's written symbol systems. That is, across the seemingly very different systems of alphabetic signs in use around the world, T and L shapes are more common than Xs, and triangular forms rarely occur (see fig. 2.2). This is not accidental, he postulates, but evidence that written signs were developed that the visual brain could easily recognize because they resembled natural forms to which it was accustomed.

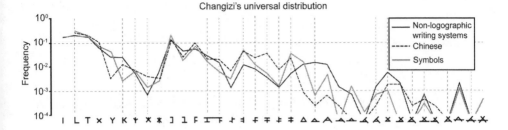

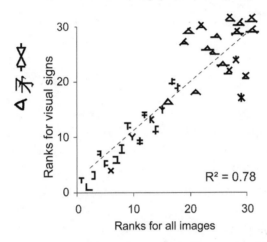

Figure 2.2. The evolutionary neurobiologist Mark Changizi has found a striking correlation between the frequency of written shapes across a range of alphabets and the recurrence of these same shapes in nature. The horizontal axis in the top graph lists these shapes and charts their frequency in 96 non-logographic writing systems (where single alphabetic characters do not represent words), Chinese, and nonlinguistic symbolic systems (traffic signs, musical notations, etc.). The three lines are strikingly similar. The bottom graph compares the frequency of human visual signs from all three of these sources (the vertical axis) to their frequency in images found in nature (horizontal axis). Again the correlations seem too close to be accidental. Reproduced by permission from Mark Changizi et al., "The Structures of Letters and Symbols Throughout Human History are Selected to Match Those Found in Objects in Natural Scenes," *American Naturalist* 167.5 (May 2006): E117–39.

If Changizi is correct, the visual features shared across the world's alphabets are based neurologically on patterns coded in the brain as a result of the long evolution of its capacities for invariant object recognition, and this finding would have important implications for linguistics and literary theory. It would not call into question Saussure's famous pronouncement, for example, that graphic signs are arbitrary, inasmuch as they are fixed by convention and acquire their significance diacritically, by their differences from other marks in the alphabetic system. But the constraints within which these contingent, culturally variable conventions are established would reflect the need of neuronal recycling to use preexisting cortical structures that are ready to be repurposed. The arbitrariness of the alphabetic sign is apparently limited by the visual geometry of familiar natural shapes, which the brain is preprogrammed to recognize. The variety of the world's alphabets testifies to the cultural contingency of the sign (any particular written code is no more necessary than any other), but their geometric similarities are evidence of the brain's preinscribed capacities for invariant visual object recognition. The sign is arbitrary and can vary because the brain is plastic and adaptable, but its variations are limited because this plasticity is constrained.

There is evidence suggesting that the linguistic sign is constrained not only by our visual apparatus but also by the workings of the auditory system. A well-known experiment shows that speakers of very different languages overwhelmingly associate a curved, rounded blob with the sound "bouba" and a sharply angled shape with "kiki."[8] This is not full-fledged Cratylism, the doctrine according to which signs are not arbitrary but are shaped by what they signify, but there may be neurobiological bases for graphic and phonemic patterns across different languages.[9] Again, as with graphic signs, the phonemic system of a language is culturally contingent because significant sounds may vary drastically from one alphabetic system to another, but the tendency of speakers of different languages to attribute a round shape to "bouba" and a sharp shape to "kiki" suggests that there may be universal, cross-cultural patterns of relating sense to sound that are based on the auditory cortex.

Although vision predominates in the recognition of graphic signs, and the reading process is triggered by activity in the letterbox area, systems associated with speaking and hearing also quickly come into play, as do areas associated with the semantics of meaning (including most importantly memory). The visual recognition of a graphic word pattern depends on associating written marks with phonemes, the significant units of sound in language, as well as with the units of meaning to which speech sounds are linked in what Saussure described as the duality of the sign.[10] As Dehaene explains, "There are two quite distinct pathways for reading"—"the direct [or semantic, visual] route, from letters to words and their meaning" and "the indirect [or phonetic, auditory] route, from letters to sounds and from sounds to meanings," a route that "fails with irregular words," however, "and with homophones such as 'too'" (40). Brain-imaging experiments show that these two reading routes—semantically oriented recognition of visual signs and phonetically oriented association of sound and sense—activate two different but connected networks in the cortex.[11]

One justifiable concern about the hypothesis of a VWFA voiced by the neuroscientists Cathy Price and Joseph Devlin is that "labelling an anatomical area with a cognitive term reinforces the notion of a one-to-one mapping between components of cognitive models and functional neuroanatomy."[12] As they rightly point out, cognition is typically more complex than this, involving multidirectional interactions among different regions of the brain. Although the VWFA is a crucial cortical node for word recognition, reading is not confined to one location in the brain; rather, it brings into relation a complex assembly of visual and auditory processes that translate letters into sounds and sounds into letters and associate signs with meaning. The reading researchers George Hruby and Usha Goswami note that even "visual word recognition is not a purely visual task" but involves "systematic correlations between visual and auditory brain areas." Similarly, they observe, "several quite distinct areas of the brain . . . are active during sounding out," and no "singular text-to-sound decoding mechanism [is] localized to a single brain area."[13] The complications of reading are for this reason a useful indicator of the dynamics of brain functioning—its "bushy" structure of overlapping, criss-crossing, back-and-forth parallel processes.

Whether sound or meaning is privileged as the route to reading is one of the biggest differences between writing systems. Not surprisingly, it is easier to learn to read in a language in which the grapheme-phoneme relation is relatively stable and transparent. Children learn to read more quickly and capably in languages with *shallow orthographies* (like Italian, Finnish, or Greek), in which phonetic and graphic structures track each other closely, than in languages (like English or Danish) in which the relation between how a word is spoken and how it is written is highly variable and often unpredictable,

Table 2.1. Relative ease of learning to read:
Percentage of errors in word reading at the end of first grade

Language	% Errors	Language	% Errors
English	66	Dutch	5
Danish	29	Italian	5
Portuguese	26	Spanish	5
French	21	German	2
Norwegian	8	Greek	2
Swedish	6		

Source: Data derived from P. H. K. Seymour, M. Aro, and J. M. Erskine, "Foundation Literacy Acquisition in European Orthographies," *British Journal of Psychology* 94.2 (2003): 143–74.

what linguists call *deep orthographies*. A comparative international study conducted by the linguist Philip Seymour showed, for example, that Danish and English schoolchildren at the end of first grade had scores that would put them in the "disability" or "nonreader" range among their peers in Italy or Finland (see table 2.1).[14]

Because reading requires the reciprocal, variable interaction of vision and sound networks, the exact relation between graphemes and phonemes in any particular language is not determined in advance by the structure of the cortex. A matter of how different sites within the brain interact, the relation between visual marks, sound patterns, and meaning units in a language system can take various forms and still allow the brain to make the necessary translations and connections between them. There are limits to what the cortex can do, however, and not all of these equivalences and interactions work equally well, and so languages with regular, predictable relations between graphemes and phonemes make reading easier than writing systems in which the deciphering of visual marks is not facilitated by sound patterns.

The differences between writing systems also cast doubt on some Darwinist explanations of the evolution of cultural phenomena like reading and writing. It is not the case that the fittest always survive in nature or in culture. Adaptations that are adequate to do the job can usually take a variety of forms, as is the case with the world's diverse languages, which have solved the problem of grapheme-phoneme translation in different, not always optimally efficient, but "good enough" ways. Noting "the mix of 'astonishing adaptive sophistication and botched improvisation' in biological design," the evolutionary critic Brian Boyd wisely cautions that "an adaptation need not be perfect to establish itself."[15] Even when evolution is a matter of genetic differences that give particular variants an advantage in sexual reproduction and thus in the replication of the feature in question, there is usually some tolerance for inefficiency, and a spectrum of solutions can adequately address the same problem. As evolutionary biologists frequently point out, nature is a tinkerer, not an engineer. To a literary theorist, evolution often seems like a massive project in what Claude Lévi-Strauss calls "bricolage," in which a structure is a result of a series of historical accidents rather than of rational design.[16] As Stephen Jay Gould succinctly declares, "The proof of evolution lies in imperfections that reveal history."[17]

Room for variation is even greater when DNA is not directly at work, as in the development and spread of national languages. Why English and Chinese, with two of the most inefficient systems of grapheme-phoneme coordination, should enjoy dominant positions among the world's languages has less to do with the sign-processing requirements of the cortex than with political and economic considerations that go beyond neuroscience. The inefficiencies of these languages are an obstacle to their spread (as any non-native speaker who has tried to learn them will attest), but they are not prohibitive or decisive (or else, wonderful thought, we would all be speaking Italian).

The ambiguities of evolution also suggest interesting and important questions about what is genetically fixed in the brain and what may vary as a result of learning, experience, and cultural differences. The neuroscientist of vision and neuroaesthetician Semir Zeki usefully distinguishes between *inherited* and *acquired* neurological traits: "An inherited brain concept, or program, is not modifiable with the acquisition of further experience throughout life," whereas "acquired brain concepts are synthetic [a result of how neurons interact] and therefore change with experience."[18] As he explains, "We are not at liberty to discard, ignore, or disobey" the brain's inherited functions and features. For example, "a normally sighted person with a normal brain . . . is not at liberty to choose not to see colors when he opens his eyes" (26). Neurological impairments provide striking evidence of the brain's inherited characteristics. Zeki notes, for example, that "a patient with damage to other visual centers in the brain, . . . those specialized for visual motion or visual form, is not impaired in color vision unless the color center itself is also damaged" (32).[19] Although the objects we become accustomed to may change with experience (no Jennifer Aniston neuron could develop without exposure to Hollywood media), the capacity to recognize visual shapes is a fixed, inherited feature of the cortex that is localized in particular areas of the brain (specialized for form, motion, color, faces, etc.). These cortical functions work automatically, without having first to be learned, and kick in again as soon as you wake up every morning, even before your first cup of coffee.

Reading is characterized by a mixture of inherited and acquired brain functions that are sometimes hard to disentangle. This mix is interestingly illustrated by some of the things that can go wrong for beginning readers. For example, a trove of experimental evidence has documented that the cognitive capacities of pigeons, dogs, and babies are characterized by *mirror symmetry*, a neurological feature with understandable evolutionary advantages.[20] It would be very useful, after all, to be able to extrapolate from one life-threatening situation to another by recognizing symmetrical features (and not having to learn the hard way that a predator coming from the left was the same as one coming from the right). Even at the cost of occasional mistakes, the ability of the brain's vision system to see symmetrical shapes as equivalent greatly aids the work of invariant object recognition.

This hard-wired neurological function can cause problems, however, especially for beginning readers, who typically confuse *b* and *d* and make other symmetry errors in the early stages of acquiring this difficult skill. These kinds of mistakes are commonly made by children learning to read in all languages. The seemingly peculiar tendency of early readers to write backwards also is not unusual and is widespread across alphabetic systems.[21] Unless a child suffers from dyslexia, this confusion is soon overcome. Nevertheless, the fact that every reader must unlearn a genetically inherited competence for mirror symmetry is striking evidence of fixed, preexisting circuits in the brain that must be disabled, overridden, or restructured in order to distinguish look-alike graphic characters. But the fact that this hard-wired competence is not the last word also shows the brain's plasticity. The confusion of beginning readers about *b/d* symmetry demonstrates both the existence of the brain's inherited cognitive features and the openness of the cortex to change and variation.

Further complications in drawing the boundary between the brain's inherited and acquired characteristics are suggested by clinical evidence from patients whose letterbox area is damaged. In one case, a young girl's epileptic seizures were so serious that they required the surgical removal of a substantial portion of the left hemisphere of her brain, including the VWFA. She eventually managed to learn to read, however, and brain imaging showed that she had done so by using undamaged areas in the right hemisphere at an exactly symmetrical location.[22] Such plasticity is unusual but not unique. As with mirror symmetry, this case shows both the strictures of the brain's inherited properties and its remarkable ability to overcome and work around their limitations. The brain is a contradictory organ with strictly defined, genetically inscribed features and a surprisingly expansive if not unlimited capacity to acquire new functions that must always take those constraints into account—but can sometimes do so in ways that suggest that they are not ultimately determining.

This contradiction is especially important for aesthetics, in particular for the question whether the brain's responses to art are universal or culturally and historically variable, a manifestation of fundamental, inherited neurological structures or evidence of the capacity of the cortex to learn, develop, and change. Zeki cites experimental evidence that suggests that from a neurological perspective beauty is both inherited and acquired. Some (but not all) scanning studies suggest that the cortical area that responds to beauty is fixed and universal: "Imaging experiments have shown that when subjects rate a painting as beautiful there is heightened activity in part of the reward system of the brain, the orbito-frontal cortex" (53). But what triggers this system varies according to what subjects regard as beautiful: "There is, in the neurobiological system, no universal Ideal of beauty, or of the form of an object, or of a landscape. Each one of these is tailored according to individual experience, and varies from one individual to the next" (47). Rather than neuroscience providing a definitive answer to the age-old question whether art is universal or relative, experimental evidence suggests that this very opposition has deep neurological foundations. Aesthetic pleasure may always be based on activity in a particular cortical location (what Zeki calls "the reward system" in the frontal lobes), but many different notions of beauty can trigger its response.

This opposition is also well illustrated by the contradictions of facial recognition, a universally inherited cortical functionality that is open to considerable individual and even cultural variation. Zeki reports that "when subjects view paintings of a particular category, for example portraits, the increase in activity within the visual brain is specific to those visual areas that have been shown to be specifically engaged when humans view faces," as opposed, say, to landscapes, which trigger "an area apparently specialized for registering places" (16–17). The cortex's reactions to faces are interestingly value laden. Romantic and maternal love activate different cortical regions, and there are "parts of the brain that are specific for faces in maternal love" (143). Brain images of "cortical deactivations . . . produced when subjects view pictures of their loved partners" reveal patterns consistent with "a suspension of judgment or a relaxation of judgmental criteria by which we assess other people, which is a function of the frontal cortex" (140). Imaging experiments show a similar effect when mothers view pictures of their children (but not the children of their friends—there the capacity for judgment remains intact).[23] As the experimental evidence shows, there are clear links between facial recognition and defined cortical areas, fine-grained even to differences between particular kinds of faces (one's child or one's beloved), and these are reenacted in aesthetic responses to their artistic representations.

How we use this cortical functionality depends on what we learn about faces, however, as other clinical evidence suggests. Oliver Sacks (himself afflicted with a neurological impairment called *prosopagnosia*, which prevents him from recognizing faces) cites clinical and experimental evidence that "there is an innate and presumably genetically determined ability to recognize faces, and this capacity gets focused in the first year or two, so that we become especially good at recognizing the sorts of faces we are likely to encounter."[24] One experimental study shows, for example, that babies at six months recognize and respond to a broad spectrum of faces, including those of other species, like monkeys, but that this range narrows, so that the response diminishes over time to kinds of faces to which the infant is not exposed (monkey faces cease to elicit a response unless this is repeatedly reinforced). Concluding that "our 'face cells,' already present at birth, need experience in order to develop fully," Sacks reflects: "The implications of this work for humans are profound. To a Chinese baby brought up in his own ethnic environment, Caucasian faces may all, relatively speaking, 'look the same,' and vice versa" (41). The behavior of the Jennifer Aniston neuron may be a striking but not unusual instance of how experience fixes and limits the visual cortex.

Other experimental evidence suggests, however, that a certain universality persists in the response to faces despite this differentiation. The neuroscientist Anjan Chatterjee reports that "cross-cultural judgments of facial beauty are quite consistent," suggesting that "the response to facial beauty is likely to be deeply encoded in our biology."[25] This claim agrees with the well-known study by the psychologists Judith Langois and Lori Roggman, which found that "attractive faces" have "average" features and that "these features seem to be perceived as attractive regardless of the racial and cultural background of the perceiver."[26] The brain's bias toward constancy and object invariance may be reflected in this averaging, even as the differences of individual experience condition the facial-recognition area to respond in particular ways to specific images. Once again neuroscientific evidence points contradictorily toward both universality and relativity.

The example of facial recognition shows that there may be a clear connection of a function to a particular inherited structure of the cortex, but how this neuronal area reacts may change as it is used, so that it becomes accustomed to responding to particular qualities, features, and forms repeatedly encountered in its environment. We may be hard wired to react in specific ways (to recognize faces, for example, and to look for similarities between them), but this wiring may in turn be genetically disposed to getting structured in a manner that reflects our personal and social history of cognitive experience. Here and elsewhere, our inherited cortical architecture seems to be universally open to individual and cultural variation.

Facial recognition is an illuminating instance of Hebb's Law, one of the foundational principles of neuroscience: "Neurons that fire together wire together."[27] Scanning studies have found, for example, that "musicians have anatomical differences in several brain areas that are involved in motor and auditory processing," with pianists, whose instruments "require precise coordination of bimanual movements," showing thicker than normal neural connections between the hemispheres of their brains.[28] A much-publicized study of London taxi drivers revealed a correlation between years of driving experience and the size of the posterior hippocampus, an area of the brain associated with navigation in birds and other animals as well as memory and conditioned fear (all understandably related to the challenges of negotiating London's streets).[29] Another experimental study showed that speakers with a command of two languages have more neuronal connections in areas of the brain associated with language use than do individuals who know only one language. These changes can occur over a limited time and then reverse themselves if the activity ceases and the second language is no longer regularly used (a finding that will not surprise anyone who learned a language as a youth but can no longer speak it).[30] Similarly, volunteers who mastered a simple juggling routine over three months of practice showed differences in scans of their motor cortex before and after their training, but these disappeared when scans were done again three months after they stopped juggling.[31] Many more examples could be cited. In all of these cases (and so not only with aesthetic judgments) the boundary between what is inherited and what is acquired is fuzzy and shifting. Certain brain regions are hard wired, but this wiring can also change.

These experiments raise the interesting question whether the brain's capacity for learning is a matter solely of its ability to make new connections between existing neurons or might also entail new cellular growth. Dehaene makes the case for plasticity through changing connections between a fixed number of neurons: "Although the number of our neurons is finite, the synapses can definitely change. Even in the adult brain, learning can still drastically alter neuronal connections" (211). Whether the number of neurons in the brain is fixed is actually an open question, however. Recent experimental evidence suggests that adults too can grow new neurons, although not everywhere in the brain. Humans apparently have a capacity for *neurogenesis* similar to what has been found in canaries, whose brains can propagate new neurons in areas used for learning songs, or in rats, whose hippocampus (one of the sites for memory) may expand in a rich learning environment.[32] If so, then there are three primary sources of plasticity in the brain: (1) new neurons may be spawned as a result of learning and memory; (2) particular neurons may develop special proclivities depending on their history of use (the Jennifer Aniston neuron); and (3) neurons wiring together as they fire together in the multidirectional interactions that can occur between different regions may result in the brain's restructuring itself after repeated experiences. The repurposing of inherited functionalities that is required by the neuronal recycling through which the brain learns to read draws on all of these kinds of plasticity.

The variability of the brain's functions makes some neuroscientists skeptical about the project of pinpointing the cortical locations of cognitive processes that the rapidly increasing sophistication of imaging technologies seems to promise to identify with ever-increasing precision. Voicing a concern of many critics who fear that a fascination with brain localization is little more than twenty-first-century phrenology, the neurophilosopher Valerie Gray Hardcastle and the neuroscientist C. Matthew Stewart warn that "brain plasticity and concomitant multifunctionality belie any serious hope of localizing functions to specific channels or areas or even modalities." Indeed, they claim, "searching for *the* function of any area is a fool's errand. The same area could be doing different things, depending on what else is happening in the rest of the brain."[33] Such radical skepticism seems too extreme, however, inasmuch as ample clinical and experimental evidence suggests that at least some core functions (having to do, for example, with basic processes of vision, like the recognition of color, shape, and motion) are localized in particular areas of the brain and are irreplaceably disabled if these regions are damaged. Nevertheless, their cautionary warnings rightly emphasize the openness of the cortex to change and development as specific neurons become accustomed to firing in response to particular signals (as in the experiments with facial recognition) and as different areas of the brain come to be connected through learning (as in the coordination of vision and sound systems in the phoneme-grapheme translations of reading). Indeed, intriguing recent fMRI evidence showing that the visual cortex of blind subjects gets repurposed for language and syntax suggests that even the core areas of the brain retain at least some plasticity.[34]

The paradoxical combination of fixity and plasticity that characterizes the cortex raises important questions about how the brain is organized, which, in turn, have crucial implications for the neuronal underpinnings of aesthetic experiences. These complications suggest some of the reasons why contemporary neuroscientists have stopped thinking of the brain as ruled by a central controller that computes information in a linear "input-output" manner.[35] As Francisco Varela explains, "Brain regions are . . . interconnected in a reciprocal fashion," and "any mental act is characterized by the concurrent participation of several functionally distinct and topographically distributed regions."[36] Referring specifically to the well-documented visual system, Ramachandran reports that "there are at least as many fibers (actually many more!) coming back from each stage of processing to an earlier stage as there are fibers going forward from each area into the next area higher up in the hierarchy."[37] If the brain can be understood on the model of high-speed computing—and such models are no longer as popular as they once were—it is more like a fluid, immensely complex, reciprocally interacting network of parallel-processing operations. As Dehaene observes, "Today, a 'bushy' vision of the brain, with several functions that operate in parallel, has replaced the earlier serial model. . . . All the brain regions operate simultaneously and in tandem, and their messages constantly crisscross each other. All the connections are also bidirectional: when a region A connects to a region B, the converse projection from B to A also exists" (64). The brain is an ensemble of simultaneously firing neurons that interact multidirectionally (bottom-up, top-down, back and forth) and get organized in a particular manner for specific tasks but can be realigned (more or less easily, depending on their physiological structure and their history) as the need and opportunity arise. No one is in charge, the system has no center, and the work gets done all the more effectively because it is distributed and interactively processed.

The complications of reading exemplify this multidirectional interactivity. Even the visual shape recognition required in reading entails too complicated a process of reciprocal interactions to program linearly. Further complexities ensue as the invariant visual form recognition operations of the letterbox area interact with the network for spoken language and as graphemes and phonemes are mutually and bidirectionally translated. These operations in turn depend on input from semantic systems in the frontal lobes that extend beyond single-word recognition. For example, how should we make sense of the homonyms *to*, *too*, and *two*, in which the sound is the same but the word form signals differences in meaning? Or a sentence like "Milk drinkers are turning to powder," or "Dealers will hear car talk at noon," or "Deaf mute gets new hearing at killing"?[38] These ambiguous sentences can only be deciphered through complicated acts of reciprocal processing, through which knowledge about semantic differences and information about the context of use must be integrated with word-form recognition in a back-and-forth, nonlinear manner. These interactions are neither exclusively top-down nor exclusively bottom-up (from higher-level meaning complexes down to grapheme-phoneme recognition, or from word-unit deciphering processes up to sentences and larger textual entities), but both down and up, back and forth, across disparate regions of the brain, reciprocally, multidirectionally, in millisecond interactions beneath conscious awareness. They are not governed by a central controller—a little man in the machine—but are complexly linked parallel processes that are organized as well as fluid, structured as well as open, patterned as well as variable.

A "bushy," decentered brain would not be likely to have aesthetic experiences located in only one area of the cortex, and this is indeed what recent imaging experiments have found. A number of neuroscientists have contested Zeki's claim that a universal "reward system" in the frontal cortex invariably fires in response to beauty. Even if he turns out to be right, that is not all that happens during aesthetic experiences (as Zeki no doubt would agree), and there is a growing consensus that the brain's responses to art are diverse and widely distributed. As the neuroscientist Marcos Nadal and his associates report, for example, "Neuroimaging studies have confirmed that there is no single brain center for aesthetic preference, and that different component processes are associated with activity in different brain regions."[39] Agreeing that "multiple components of art defy functional localization in the brain," the neuroaesthetician Dahlia Zaidel reports that brain scans have found "multiple regions to be engaged and no specific regions to be associated with the specific categories of ugly, beautiful, or neutral."[40] This is also true, according to the neuroscientist Oshin Vartanian, of various emotions often associated with art: "No dissociable activation pattern can be mapped out for discrete emotions such as fear, anger, happiness, or sadness" either when they are experienced in life or when they are aesthetically reenacted.[41]

If it is a mistake to search for the unique cortical location of aesthetic experience, it might make sense to try an approach more consistent with the decentered, multidirectionally and reciprocally interacting structure of the brain. This is the kind of neurological organization that one might expect to be able to support aesthetic experiences of both harmony and dissonance. Although looking for where the brain experiences art is wrongheaded, it might be productive to ask how experiences of harmony and dissonance activate multidirectional, reciprocal interactions between regions of the cortex and, further, to inquire into what the purposes of such processes might be.

Different kinds of aesthetic harmony would no doubt have different neuronal correlates. Harmonies can be pleasurable either because they resonate with recognizable patterns or because they suggest new relations. Some harmonies may activate particular modes of integrative, reciprocal interaction that are familiar from past patterns of neuronal processing, while others may reconfigure relations between cortical areas in new ways. The brain's ability as a fluid parallel-processing system to organize and reorganize itself could be responsive to both kinds of harmonic configuration.

Harmony is not uniformity or homogeneity but, rather, a structure of interrelated differences, and harmonies that develop and change over time achieve their effects by performing transformations in these relations through adjustments and alterations that facilitate integration. Unlike the disturbances of noise, even surprising disruptions along the way can serve integrative purposes, such as establishing structures of foregrounding and backgrounding or calling attention to a particular shape or pattern. Noting that "in music, the existence of novel, unexpected sound events is in a key position to keep the listener alert" and is "also important in creating positive emotions," Mari

Tervaniemi reports interesting brain-imaging experiments charting the responses of human subjects to unexpected or anomalous musical sounds: "The chords most strongly violating the subjects' expectancies evoked earliest and largest responses."[42] The use of surprise to focus or redirect attention is a feature not only of dissonant but also of predominantly harmonious art, and that is because harmony is a structure of differences open to change and variation. Harmonious art is a paradoxical system of similarities constructed out of differences, and this paradox plays with the brain's contradictory characteristics as a parallel-processing network that operates through simultaneous reciprocal interactions among different cortical regions.

The neurobiology of *consonance* in music, a particular kind of aural harmony, is beset by interesting complications that illustrate this paradox and illuminate the general workings of aesthetic harmony. Daniel Levitin reports that "a great deal of research has focused on the problem of why we find consonant some intervals and not others, and there is currently no agreement about this."[43] As he explains, "The brain stem and the dorsal cochlear nucleus—structures so primitive that all vertebrates have them—can distinguish between consonance and dissonance; this distinction happens before the higher level, human brain region—the cortex—gets involved." But he also points out that the apparent universality of the distinction between consonance and dissonance is culturally relative as well: "Two notes can sound dissonant together, both when played simultaneously or in sequence, if the sequence does not conform to the customs we have learned that go with our musical idioms" (74, 75). Furthermore, the neuroscientist of music Aniruddh D. Patel observes that "there may be individual differences in the degree to which people prefer consonant or dissonant intervals." He points out that "there are cultures in which rough-sounding intervals . . . are considered highly pleasing, as in certain types of polyphonic vocal music in Bulgaria."[44] He also reports, however, that infants develop a preference for consonant sound combinations at as early as two months, whereas experiments with macaque monkeys show that they can distinguish consonance from dissonance but do not prefer one or the other (they do not reposition themselves to avoid dissonance coming from one end of a V-shaped maze, although they will move away from loud sounds toward softer ones) (see 396–97). Consonance may be a cognitive universal, but it is also relative to individuals, cultures, and species.

As a structure of differences, harmony may be based on fundamental cognitive capacities for recognizing certain kinds of distinctions (in music, the auditory system's ability to process different frequencies and to recognize regularities in their relationships that music theorists identify as consonance). But patterns of difference can be subject to wide personal and cultural variation. Differences that seem harmonious, or "consonant," to one individual or group can seem dissonant to another, and some individuals and groups may come to prefer a roughness or atonality that others would find disruptive and unpleasant (or, as in the case of some monkeys, may not particularly care about one way or the other). These complications can occur only because harmony entails both similarity and difference—only because, indeed, its constitutive similarities are patterns of difference.

These contradictions recur in the well-known debate over whether atonal music is unnatural, intrinsically antagonistic to our perceptual capacities, or is an acquired taste just as tonal music is. Theodor W. Adorno famously argues, for example, that Schoenberg's twelve-tone scale meets resistance primarily because we are not accustomed to it and are blinded to other musical possibilities by our ideologically conservative preference for harmonic closure: "New music . . . takes up a social stance . . . in that it abandons the deception of harmony," and "the shocks of the incomprehensible . . . illuminate the meaningless world."[45] This is a message we reject for political reasons, he argues, but there is nothing inevitable about our resistance. Writing about neuroscience and music, Jonah Lehrer concurs: "Our sense of sound is a work in progress," and "nothing is difficult forever."[46]

The experimental evidence suggests, however, that there may be limits to what the auditory system can assimilate, at least at this evolutionary stage. The psychologist of music Sandra Trehub cites evidence that "the central nervous system, acting in conjunction with motor systems, predisposes us to perceive certain pitch relationships, temporal proportions, and melodic structures as well shaped and stable," and she consequently concludes that "avant-garde composers, by abandoning the tonal system, . . . may have created musical forms that are inherently difficult for human listeners, forms requiring deliberate, effortful learning for their understanding and appreciation."[47] Some dissonant forms may be more than the auditory system's integrative capacities can handle, and some art may be inherently resistant to assimilation. The reasons for this resistance may be as much biological as political. No matter how often we hear Schoenberg, there will be something difficult about the experience because of the neurobiology of the comprehension of musical sound.

Here again the brain is plastic, but there are limits to its adaptability. Art is a matter of learned conventions, and some works that seem odd, difficult, and unnatural may become less so as we become accustomed to them and learn how to understand their strategies—how to recognize the patterns that make their disruptions meaningful, the principles and purposes behind their breaking of the rules. But the materials that the conventions of art manipulate (as with the graphic shapes used in various alphabetic systems) are constrained by fundamental structures and processes of the brain. Different patterns of harmonic or dissonant relationship facilitate, test, challenge, or strain the brain's capacities for integration in different ways, and some patterns are intrinsically easier or harder for us to assimilate.

For example, turning from music to literature, although nineteenth-century Paris is farther away from my students' experience than late twentieth-century Southern California, they still find Balzac easier to comprehend than Thomas Pynchon because Balzac's authoritative narrator's assurances that "all is true" and that everything he tells us will fit together in coherent patterns facilitate integration more than the paradoxical, paranoid mysteries that baffle Oedipa Maas in *The Crying of Lot 49* (although the joys of this novel's oddball humor are for many students a more than ample compensation). Such differences remain even for expert readers (like myself). For example, no matter how often I read *Ulysses*, it remains more difficult and resistant to comprehension than some novels in the realistic tradition (by, say, Jane Austen, Charles Dickens, or George Eliot) that I nevertheless also value highly—novels that have their own subtleties and complexities, to be sure, but that facilitate integration more than puckish, rebellious James Joyce does. Samuel Beckett, Alain Robbe-Grillet, and David Foster Wallace push to the limits—and beyond—the brain's powers of integration even for students and critics well versed in the conventions of avant-garde writing. These differences are a result of how writers manipulate contingent, cultural conventions, and in doing so they can teach us to read in new ways (what was unnatural can become recognizable and familiar), but some experiments with form may be simply too much for the brain to handle smoothly and routinely.

If harmony is pleasing for its own sake (consistent with an art-for-art's-sake aesthetic), one might expect to find this reinforced by the brain's chemistry. The *neuromodulator hormones*, which induce feelings of pleasure and encourage connection with others, may be released when the symmetries and reciprocal resonances of art activate the brain's parallel-processing networks. Harmony in art is often associated with love, and there is considerable experimental evidence that powerful cortical opiates are released by the experience of love. According to Zeki, "Most brain regions, including subcortical regions, that have been determined to contain receptors for oxytocin and vasopressin [two of the major neuromodulator hormones] are activated by both romantic and maternal love [i.e., of a mother for her children]" (145). He admits that his comments about the effects of these hormones on the brain are speculative, based on experimental evidence with prairie voles (a mammalian species that, atypically, is monogamous because of the role of oxytocin and vasopressin in its mating cycle), which may or may not equate to humans.[48] Art is not reducible to hormones, and harmony is not the same as love, but it would be surprising if the often-reported mood-altering effects of harmonious aesthetic experiences were not reflected in and reinforced by the brain's chemistry.

If, as the well-known Horatian maxim states, art both "pleases and instructs," there are probably also neurological bases for the pleasures of harmony that have to do with the learning it makes possible. Such learning may happen through the invocation of familiar patterns (the precedents provided by classical models) that extend and strengthen already existing cross-cortical processing networks. Or it may be facilitated by the introduction of new structures that reconfigure the brain's connections and bring different cortical regions into new relationships, rewiring synapses to reshape the brain's plasticity in new forms of consonance. If harmony alone is not sufficient to qualify an experience as aesthetic, however, that is because the symmetries and resonances of cortical parallel processing can manifest themselves in many different kinds of phenomena. As Martin Skov and Oshin Vartanian remind us, "Aesthetic processes typify our interactions with a much wider array of objects than artworks alone," and "the component neural processes involved in producing or appreciating art themselves are common, not unique."[49] This is certainly true of harmony. For that very reason, however, the ways in which art's harmonies reinforce or rewire neuronal assemblies may have an impact on modes of perception and cognition in everyday life, and the neurological workings of the aesthetic experience may provide clues to the larger mystery of reciprocal interaction in the brain.

The artistic values of disruption and disjunction are also consistent with a decentered, parallel-processing model of the brain. Aesthetically pleasing, purposive dissonance is not noise. Unlike the randomness and disorganization of noise, aesthetically meaningful dissonance is an internally coherent structure of differences that is strategically opposed to the harmonies it disrupts.[50] The cognitive purposes of dissonance have to do with how the plasticity of the brain organizes itself. Long before contemporary imaging technologies gave deep insights into the internal architecture and functioning of the cortex, the psychologist William James intuited the brain's paradoxical combination of variability and fixity in his well-known observations on habit: "The whole plasticity of the brain sums itself up in two words when we call it an organ in which currents pouring in from the sense-organs make with extreme facility paths which do not easily disappear."[51] According to James's famous analysis, habit is a double-edged sword. All too easily formed by repeated experiences, habits make our actions more efficient by eliminating the need for conscious, purposive deliberation, but they then lock us into patterns that are hard to break (see 1:112–14, 121–27). A brain that is a decentered ensemble of multidirectional interactions needs to establish patterns to function effectively (or at all), but this gain comes at the cost of a loss in the flexibility and openness to new connections that are responsible for its remarkable responsiveness and adaptability.

A brain that is both plastic and structured stands to benefit from aesthetic dissonances. The negative consequences of habit are the target of the Russian Formalist Viktor Shklovsky's well-known aesthetic of defamiliarization:

> Habitualization devours works, clothes, furniture, one's wife, and the fear of war. "If the whole complex lives of many people go on unconsciously, then such lives are as if they had never been." And art exists that one may recover the sensation of life; it exists to make one feel things, to make the stone *stony*. The purpose of art is to impart the sensation of things as they are perceived and not as they are known. The technique of art is to make objects "unfamiliar," to make forms difficult, to increase the difficulty and length of perception because the process of perception is an aesthetic end in itself and must be prolonged.[52]

The dissonances of defamiliarization, according to this view, have the value of reviving the brain's responsiveness, its elasticity, against the tendency of particular neurons and specific cortical connections to become fixed through repeated use.

This was the purpose, according to Jonah Lehrer, of "the sweeping dissonance" of Stravinsky's *Rite of Spring*, which was first perceived as "a monstrous migraine of sound"; "orchestral music had become boring," and so "Stravinsky anticipated the anticipations of his audience and then refused them every single one" (121, 125, 132). Lehrer exaggerates, however, when he claims that "Stravinsky's music is *all* violation. . . . Dissonance never submits to consonance" (132, emphasis in original). Dissonant art defamiliarizes by violating a norm, and this requires the invocation of the norm, the structure that is laid bare by being transgressed, opposed, and overturned. Further, these acts of violation are themselves meaningful patterns of contestation and reconfiguration, which is why they are then susceptible to routinization and habitualization, so much so that they may become conventional and even banal (as when, Lehrer notes, Stravinsky is adapted by Walt Disney's *Fantasia* [see 139]). They may then in turn require defamiliarization to regain their aesthetic and cognitive power. Again, the dissonances and disruptions of some avant-garde art may be more resistant to assimilation than those of other experimental works, and Disney's cooptation of Stravinsky may support Adorno's assessment that Schoenberg is the more daring, radical composer (see 155–58). But great works of art are not immune to the dangers of habitualization.

The very familiarity of well-known art may not always breed contempt (although it can, as when an avant-garde artist paints a mustache on the *Mona Lisa*), but it has often been remarked that repeated exposure will almost certainly diminish a work's aesthetic effect (as in the case of this smile, which has become a cliché). Adorno notes, for example, that "adequately listening to the same Beethoven works that the fellow in the subway contentedly whistles in fact requires far greater effort than does adequately listening to the most avant-garde music" (12). Other examples spring readily to mind, such as Vivaldi's "Four Seasons" or any of the classical pieces one hears in the elevator or while waiting on hold on the telephone. As the phenomenological theorist Hans Robert Jauss points out, the "classical" status of "the so-called masterworks especially" risks dulling their effect, because "their beautiful form . . . has become self-evident," so that "it requires a special effort to read them 'against the grain' . . . to catch sight of their artistic character once again."[53] Paradoxically, one of the worst things that can happen to a work of art is for it to become canonical, and rendering classical works dissonant may be necessary to revivify their harmonies. This is often the intent of the radical restagings of well-known plays or operas, which can outrage an audience, and this may be exactly what the director has in mind. It is then an interesting question whether such protests show that the defamiliarization has successfully brought into focus what repeated performance had dulled or that the dissonances entailed in reading against the grain have overwhelmed the harmonies that constitute the work's value.

As these ambiguities demonstrate, the opposition between aesthetics of harmony and dissonance is based neurobiologically on the values and pleasures of making neuronal connections and of breaking their grip to enable new modes of cortical organization. That is what is going on neurologically when the dissonances of defamiliarizing art upset habitual modes of synthesis and what was dulled becomes visible again as new configurations of meaning, based on new neuronal assemblies, emerge. Debates about whether experimental strategies of defamiliarization are aesthetically successful are, at another level, arguments about whether the restagings in question have facilitated this neuronal restructuring. That cannot be decided in advance, and, further, it will vary for different recipients with different habits and degrees of cortical flexibility.

There is interesting experimental evidence that the making and breaking of neuronal connections stimulates the expression of neurotransmitters strongly associated with pleasure in ways that no doubt affect aesthetic experiences. The neuroscience of pleasure is a complicated and still evolving area of research, and generalizations about its workings should be ventured carefully.[54] In addition to oxytocin and vasopressin, which I discussed above in the analysis of harmony and love, the neurotransmitter dopamine apparently also plays a crucial role. As the neuroscientist David J. Linden points out, "Dopamine-containing neurons of the ventral tegmental area (VTA) . . . send dopamine-releasing axons to other brain areas," including the amygdala (associated with emotions), the dorsal striatum (involved in habit formation and learning), the hippocampus (a memory center), and the prefrontal cortex (activated by judgment and planning) (16). Experiences that stimulate the VTA release dopamine and are felt as pleasurable, and the VTA network would seem to link an extraordinarily wide range of cortical functions. One reason why dopamine has received considerable attention is that its release is also stimulated by amphetamines, nicotine, heroin, and cocaine and is implicated in the neurobiology of addiction. Art is not, of course, strongly addictive as the action of these drugs is (although the similar phenomenon that repeated experiences of a work are dulling may suggest parallels worth exploring), but the release of dopamine in the assembly and disassembly of neuronal connections across such a broad array of brain functions is certainly part of why aesthetic experience gives pleasure.

The way that *opioid receptors* are structured in the brain facilitates both consistency building and responsiveness to novelty, once again with implications for aesthetic experience. Noting that "the brain is wired for pleasure," Irving Biederman and Edward A. Vessel report that the number and density of these receptors increases in the rear visual cortex in a "so-called association area," where interpretive connections are made and "visual information engages our memories."[55] A similar gradient structure has been found in the auditory system of the macaque monkey (253). Neurotransmitters are chemical modulators that only reinforce brain activity and do not themselves cause any behavior they influence. Nevertheless, the increased concentration of receptors for such chemicals in areas where the brain connects perceptions with memories is no doubt "key to the pleasures we derive from acquiring new information," as Biederman and Vessel argue, and helps to account, "at least partially, for the human preference for experiences that are both novel . . . and richly interpretable" (249, 250). Both harmony and dissonance are, of course, experiences of this kind. Although aesthetic joy is more than a chemical high, the gradient structure of the brain's receptors for neurotransmitters is consistent with the pleasures associated both with art's beautiful syntheses and with its surprising disruptions. There are neurochemical reasons, then, for why art both pleases and instructs.

The brain's ability to go back and forth between harmony and dissonance is evidence of a fundamental playfulness characteristic of high-level cortical functioning. It is not accidental, I think, that theorists from Kant and Schiller to Gadamer and Iser have associated the aesthetic experience with play. According to Kant, for example, the enjoyment of beauty is linked to the "free play of the cognitive faculties" that the disinterested contemplation of art makes possible.[56] Although Kant's notion of play is closely tied to his particular conception of art as noninstrumental, depragmatized "purposiveness without a purpose," the phenomenological theorists Gadamer and Iser propose a theory of play as a "to-and-fro" movement that can take a variety of forms, associated with a disparate range of aesthetic theories and experiences. Not all play, even in art, serves Kantian disinterestedness.

Iser distinguishes, for example, between instrumental games, which seek to fix meaning by establishing a winner and a loser, and open-ended forms of play, which resist closure and try to keep differences in motion without ever finally settling down. Analogously, as he points out, some works of art seek closure (the autonomy of self-referential poetic form that Jakobson emphasizes, for instance, and that the New Critics valued), while others resist determinant interpretation (the inconclusiveness of ambiguous works and open forms that post-structuralist and postmodern critics prefer). Like games that seek to fix meaning and determine a result, some works of fiction pretend to offer a realistic representation of a world that claims to be "there," with the specificity and reliability of everyday life, while others multiply mysteries and indeterminacies and keep the reader guessing to the end and beyond. These are only some of the many oppositions that can characterize aesthetic play. What Iser calls the "ever-decentering movement," the continual "oscillation, or to-and-fro movement, [that] is basic to play," is inherently variable and open to a variety of aesthetic manifestations in modes of interaction characterized by different rules, boundaries, structures, and aims.[57] Different genres, different periods, and different works of art can be defined by the various kinds of aesthetic play they value and promote.

The ability to play in all of these different ways is made possible by the brain's contradictory, decentered structure. The duality of the brain's organization, its dependence on established patterns and its openness to new combinations, can support a variety of games, ranging from closure to open-endedness. Different kinds of play structure the brain's fixity and plasticity in different ways, and our capacity to take aesthetic pleasure in different modes of play is a reflection of this fundamental, neuronally based variability. A brain that was more fixed or more chaotic would not be able to engage in different games or have different aesthetic experiences, because it would be either too closed to play or too open. The fundamental duality of the brain's organization, its need not only for coordination and connections to make sense of the world but also for elasticity and adaptability to reconfigure itself in light of new challenges, makes it possible and useful for it to play.

The decentered, to-and-fro movement of play requires a decentered brain. One purpose of art no doubt is to give us experiences of playing in different ways, some more harmonious, some more dissonant, and thereby to provide us with opportunities to alternate between cognitive modes (recognizing that cognition is embodied and also includes emotions). Although one should be careful about making speculative Darwinian arguments about evolutionary advantages, which are easy to conjecture but difficult to prove, it is hard to imagine that the ability to play in a variety of ways is not beneficial to the cortical organization and resiliency of our species and therefore to its survival. The same argument can be made for any of the many other species that engage in playful activity.[58] If I may speak metaphorically, the brain "likes" to play (and it has been experimentally documented, indeed, that animal and human play releases the pleasure-related neurotransmitter dopamine).[59] And playing is good for it. If art in all of its multiple forms has neurological value, one reason is that it facilitates the playfulness of the brain.

CHAPTER 3

The Neuroscience of the Hermeneutic Circle

One of the curiosities of contemporary neuroscience is that it has redis-covered some of the ancient truths of hermeneutics, the long philo-sophical tradition devoted to the study of interpretation.[1] The central tenet of hermeneutics is that interpretation is circular. The *hermeneutic circle* stipu-lates that one can understand a text (or any state of affairs) only by grasping in advance the relation of a specific part to the whole in which it belongs, even if one can only arrive at a sense of the whole by working through its parts. This epistemological paradox is not consistent with a linear model of the brain as a scanning machine or an "input-output" computer. It is, how-ever, completely congruent with a "bushy" model of the brain as a decen-tered, multidirectional ensemble of parallel-processing operations.

The circularity of interpretation is evident every time we read. Hence the phenomenological theorist Wolfgang Iser's description of reading as an an-ticipatory and retrospective process of building consistency and constructing patterns.[2] Reading is not a linear process of adding sign to sign, scanning one after the other in a sequential manner. Rather, reading a text requires the rec-ognition of patterns, and a pattern is a reciprocal construction of an overall order and its constituent parts, the overarching arrangement making sense of the details by their relation to one another, even as their configuration only emerges as its parts fit together.

We put the hermeneutic circle into motion whenever we attribute mean-ing to a text. For example, the genre of a text, the kind of linguistic artifact we think we are reading, is an important guide for interpretation because it provides an anticipatory understanding of the whole into which we should fit its details. I interpret the sentence "Joe was murdered" differently if it occurs in a newspaper than I do if it occurs in a novel. The kind of text in which a sentence occurs provides a frame through which we see it. Whether readers regard Henry James's novella *Turn of the Screw* as a ghost story or as a psychological study of mental distress will determine how they understand many details of the story—for example, whether the "ghosts" the governess sees are "real" or hallucinations.[3] When readers disagree about the meaning of a text, the reason is often that they think they are dealing with different kinds of linguistic artifact, and so they are guided by different expectations in their quest for patterns.

Other indications of context—the historical period in which a text was produced, for example, or what we already know about the author's typical themes, interests, and style of writing—are similarly valuable clues for understanding because they suggest hints about how details in a text will fit together. Trying to make sense of a poem without knowing when it was written or by whom can be extremely difficult, because one is deprived of likely hypotheses to test about how to connect its parts. Learning the artistic conventions that prevailed during a particular period or having an acquaintance with an author's other works may not provide automatic comprehension of a text, because the construction of meaning still requires to-and-fro adjustments between the details of the work and the configurations one expects (and may or may not find). But without clues of this sort about the probable overarching design of the text, an interpreter doesn't know how to begin.

The configurative, recursive, to-and-fro movement of this circle is visible in the well-known perceptual shifts that can occur with those ambiguous figures that first seem like a duck but may change into a rabbit, or an urn that may transform into two faces (see fig. 3.1). In a circular manner, our reading of the beak of the duck shifts if we can be induced to see the figure instead as a rabbit, in which case this shape changes meaning and becomes a pair of ears. Similarly, two squiggly vertical lines may seem to compose the contours of an urn until we shift the pattern to which we assume they belong and consequently construe its details differently, with two juxtaposed mouths, noses, and foreheads now facing one another across the empty space that had previously formed the urn's body. Gestalt shifts enact the recursive, circular interdependence of part and whole in interpretation and show that the hermeneutic circle characterizes not only how we read texts but, more fundamentally, how we understand the world. These epistemological connections explain why ambiguous figures are so fascinating to aestheticians interested in visual representation, like E. H. Gombrich, or to neuroscientists seeking to unlock the mysteries of vision, like Semir Zeki. The intuition that ambiguous

A. Rabbit or Duck?   B. Urn or Two Faces?

Figure 3.1. Ambiguous figures. Drawings by Maggie Buck Armstrong.

figures can yield fundamental insights about how we know the world is no doubt also one reason why Zeki's work on vision led him to neuroaesthetics and especially to the role of ambiguity in art.[4]

These examples suggest that the circle metaphor is misleading, however, because the recursive, to-and-fro process of configuring parts and wholes is temporal, not spatial, and does not go round and round in the same path but moves back and forth, shifting and rearranging itself. The term *hermeneutic spiral* would perhaps be more accurate. The to-and-fro movements of interpretation can make it an unpredictable activity—an important example of the brain's proclivity to "play." Martin Heidegger emphasizes the hermeneutic circle's temporality when he argues that interpretation is inherently futural, always guided by an anticipatory understanding that projects a range of meanings that the state of affairs in question may have, what he calls the "forestructure" (*Vorstruktur*) of interpretation.[5] To interpret is therefore a process of "catch-up," he argues, as we explicate (*aus-legen*, or "lay out") possibilities for which our expectations have paved the way. We are always ahead of ourselves when we understand, especially (but not only) when we read.

Hence the importance of surprise—those disorienting experiences of reversal when parts refuse to fit (Joe isn't dead after all!) and we must reorient ourselves by reconfiguring the patterns into which we had aligned them (the report of his murder was a hoax, someone playing a joke on us, or simply an error, a case of mistaken identity, or a fictional story we had wrongly taken for a factual report, among other possibilities). We would not experience surprise if interpretation did not depend on expectations, our anticipatory understanding of part-whole relations. Setting the reader up to be surprised is a classic element of the storyteller's art and a key to the construction of narrative.[6] The twists and turns of a story solicit our expectations in order either to confirm or, more likely, overturn them—experienced readers paradoxically expect to be surprised, the tantalizing question being how and when—and narrators can play with readers in this way only because reading is a temporal process of anticipating how parts will fit together. The reconfiguring and realignment of what we have read that must take place after we have been surprised demonstrates the reverse movement of the "spiral" as we rearrange parts into a different whole than we had thought we were constructing.

The role of expectations in understanding suggests two issues that have been of central concern to both hermeneutic theory and neuroscience and that will guide this chapter's explorations of the neuroscientific foundations and implications of the hermeneutic circle. First, is the circularity of interpretation potentially a vicious circle? In hermeneutic theory, the issue here is that our expectations may be self-fulfilling, inasmuch as what we see in a text may simply be produced by the patterns we project in a mutually confirming manner. Our interpretive constructs may be rigidly self-reinforcing, making us resistant, blind, or indifferent to any need for change, alteration, or adjustment. This dilemma gives rise to important debates about validity in interpretation.[7] Surprise—anomalies not fitting expected patterns—is one safeguard against vicious circularity. But why are we ever surprised, and are we always surprised when we should be?

In neuroscience, similar questions arise about the conflict between the value of constancy and stability in perception and the need for flexibility and openness. Patterns are useful, even essential, in order for all of our sensory systems—not only vision but also hearing, smell, taste, and even touch—to provide stable constructions of data from an outside world perpetually in flux. But how can these systems recognize situations in which the very instabilities and discontinuities that they regularize (and thereby suppress) instead call for shifts in the patterns into which we construct them? Ambiguous states of affairs are especially illuminating to neuroscience because they lay bare the processes by which pattern and flexibility work out their competing imperatives.[8]

The second issue has to do with the possibility of conflicting interpretations to which so-called multi-stable images like ambiguous figures may give rise. The capacity of such images to stabilize in multiple, incommensurable patterns—rabbit or duck? urn or two faces?—is a special case of the general hermeneutic phenomenon of interpretive conflict. Ambiguous texts like *The Turn of the Screw* are especially vivid instances of hermeneutic disagreement: is the governess acutely sensitive to the evil threatening her wards, or is she an insane hysteric whose fantasies are the real danger (and the cause of one child's death)? But such conflicts over meaning are not unique to ambiguous texts. Like the rabbit-duck figure, an ambiguous novel like *The Turn of the Screw* calls attention to epistemological processes also at play, but less visible, in other instances. This novella's ambiguity foregrounds possibilities of interpretive disagreement that can also occur with less self-consciously elusive works.

The material, neurobiological bases of these hermeneutic processes may emerge more clearly if we examine in some detail how vision works. There are several good reasons for focusing on the visual system. For one thing, it is the most studied of the brain's systems, and it occupies a larger proportion of the cortex than any other sensory system, in part because of the importance of sight for survival. Further, the neuronal recycling through which reading repurposes various preexisting cortical structures grafts itself onto the visual system, and one would consequently expect that findings about vision would offer insights into reading. (We have already seen, for example, the relevance of invariant object recognition for the construction of letters in an alphabet.) Finally, neuroscientists have found that the fundamental principles of the visual system pertain as well to hearing and touch.[9] Despite differences in the receptors from which signals originate in the eyes, ears, and skin, there are basic similarities in how these signals are processed in the brain: differential inputs from receptors (patterns of excitation-inhibition like an on-off switch) mapping onto topographical areas of the cortex, parallel processing of these signals in different cortical areas, and feedforward-feedback interactions between them through reciprocal neuronal connections. These functional similarities are not surprising, because hermeneutic circularity characterizes not only vision and reading but also hearing and even touch—all depend on the recognition of patterns and the configuration of gestalts—and so it would make sense that their underlying neurological processes would have much in common.

Vision is inherently hermeneutic. As the neuroscientist of vision Margaret Livingstone points out, "Vision is information processing, not image transmission," providing "information about what is out there in the world, and how to act on it—not a picture to be looked at."[10] Despite centuries of visual metaphors for knowledge that depict the mind as a mirror, the sensation we have that we are watching a full-color picture that corresponds point by point with the external world is an illusion—a complex illusion that the brain constructs so efficiently that we rarely notice the hermeneutic machinery that produces it.[11] As Semir Zeki notes, however, "What we see is determined as much by the organisation and laws of the brain as by the physical reality of the external world."[12] That does not mean that the hermeneutic constructions of vision are merely fictional creations, solely of our own making. As the neuroscientist Zoltán Jakab observes, "Perception need not be veridical in every respect, in order to be adaptive. For an organism with color vision, the important thing is only that color perception enhances visual surface discrimination, that it helps to pick out red or orange berries from among green leaves and the like."[13] Color does not exist as such in the external world but is a complex construction of constancies out of a flux of inputs that well illustrates how the brain creates regular patterns for pragmatic purposes. These processes, like the hermeneutic circle, are both responsive and creative, reacting to signals that they in turn manipulate and structure in a to-and-fro, interactive manner.

The hermeneutic processes of vision have evolved and survived not because they perfectly mirror the external world but because of their pragmatic usefulness in negotiating our way through it.[14] According to Livingstone, "The function of our visual systems is not simply to reproduce the pattern of light falling on the retina (so that someone inside our heads can look at the picture) but to extract biologically important information from our environment" (90). Vision begins when light hits receptors on the back wall of the retina, and these in turn send signals through a complex pathway to an area known as "V1," at the rear of the cortex. But as Zeki explains, even at this early stage, before these signals have been fully processed, the "retinal map" in V1, "unlike an ordinary photographic plate, is not a straightforward, undeformed, translation" of the world: "the centre of the retina, known as the fovea, which has the highest density of receptors and which we use when we want to fixate objects and study them in detail, is given a disproportionately large amount of cortex; the peripheral part of the retina, by contrast, is under-emphasised relative to its retinal extent" (*Inner Vision* 17). Even the initial translation of light into optical signals in the retina is not a one-to-one "mirror" but an interpretation, because, as Zeki observes, "it is a map that emphasises a particular part of the field of view" (17), and the representation of retinal signals in V1 reproduces this distortion of the visual scene. This is an illuminating distortion, however, because it gives us a perspective on the world. One need not be a reader of Henry James to realize that an awareness of point of view has a variety of pragmatic uses (it is useful to know, for example, the direction your predator is coming from), and visual structures that provide perspectively oriented interpretations of the world have many significant advantages over point-to-point mirroring.

Things quickly become more complicated as these signals are sent from V1 to adjacent areas of the rear visual cortex, which are functionally specialized to detect orientation, motion, and color (V2–V5) and to identify objects and faces (see fig. 3.2). As Zeki explains, "The brain handles different attributes of the visual scene in different, geographically distinct, subdivisions," and "vision is therefore organised according to a parallel, modular system" (58–59). For example, because color is decoded in the V4 area and motion is detected in V5, injuries to those sites cause different visual disabilities, and "a lesion in one area does not invade and disable the perceptual territory of the other." Further, the processing in these different areas is asynchronous. Although "over longer periods of time, in excess of 500 milliseconds [0.5 seconds], we do see different attributes in perfect temporal and spatial registration (the attributes are 'bound' together)," at smaller intervals the different processes are disjunctive: "color is perceived before motion by ~80 milliseconds [0.08 seconds]," and "locations are perceived before colors, which are perceived before orientations."[15] Although we do not notice these discrepancies, they demonstrate that vision is a complex process of discrimination and combination, a binding (to use the customary neuroscientific term) of separate processes in different, relatively autonomous regions of the visual cortex.

This synthesis occurs, further, without a central controller or little man in the machine to guide it. As Zeki explains, "There is no single master area to which all the visual areas uniquely project." "Vision," he writes, "consists of many micro-conscious events, each one tied to the activity of a given station in a processing system. A conscious experience does not depend upon a final stage, precisely because there is no final stage in the cortex" (*Inner Vision* 71, 73). We have meaningful visual experiences not because a homunculus in the brain watches and interprets the images projected on the screen by the eye's camera (there is no such thing) but, rather, as a result of reciprocal, to-and-fro interactions between distinct but interconnected cortical areas that happen too quickly for us to be aware of them. Such interactions could not

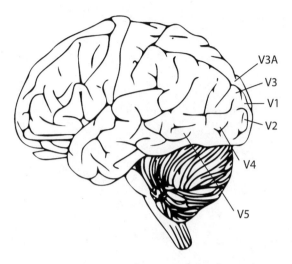

Figure 3.2. Approximate location of visual areas in the cortex. V1 registers visual stimuli, V2 responds to shapes, V3/3A is sensitive to orientation, V4 responds to color, and V5 detects motion. Drawing by Maggie Buck Armstrong.

take place and would not be necessary if cortical processing were perfectly simultaneous and homogeneous. Instead, the temporal and spatial disjunctions that divide these related areas make "play" between them both possible and essential for the production of coherent visual interpretations of the world.

The differentiation of signals that are integrated in visual experience begins in the structure of the eye. As is well known, the eye features two kinds of photoreceptors: *rods*, which are sensitive to very dim light, and *cones* of three sorts, which construct color by their responsiveness to different wavelengths in the spectrum but which do not react in nightlike conditions (so that we are color blind in the dark).[16] There is a further, less well known division between two kinds of *ganglion cells*, which receive signals from these photoreceptors and transport them to the cortex (axons of the ganglion cells project out of the eye into the cortex to form the optic nerve). The retina contains more than 100 million photoreceptors, but only 1 million axons carry signals out of the eye, and that is because two kinds of ganglion cells collect inputs on different scales. As Livingstone explains, the large ganglion cells, with broader receptive fields, convey information about "motion, space, position, depth (three-dimensionality), figure/ground segregation, and the overall organization of the visual scene"; neuroscientists call this "the 'Where' system" (50). The smaller, more fine grained ganglion cells—some in the foveal area "so tiny that they receive input from only a single photoreceptor"—are "responsible for our ability to recognize objects, including faces, in color and in complex detail" in what is referred to as "the 'What' system" (49, 50). The "Where" system is older in evolutionary terms and is common to all mammals, whereas the "What" system is well developed only among humans and other primates, although it may exist in more rudimentary forms. "Lower mammals are much less sensitive to color than we are, and they are not able to scrutinize objects and accurately discriminate them on the basis of visual attributes. Instead they are sensitive to things that move, because things that move—either prey or predator—are likely to be important" (52).[17]

Wondering why the "What" system evolved separately, as "a primate add-on," Livingstone speculates: "As the more complicated primate visual system evolved, the original system was maintained, probably because it was simpler to overlay color vision and object recognition onto the existing system than it would have been to incorporate the two" (52). As she explains, there are also pragmatic hermeneutic reasons for this "segregation of our visual system," inasmuch as "it is more efficient to carry information about—and make calculations about—an object's appearance (its shape and color) separately from information about its position and trajectory" (52). As she notes, the technology of high-definition television similarly breaks images down into differentiated packets of information about color, shape, and location in order to compress its signal to fit the limits of the available bandwidth (see 194–95). Unless something goes wrong and the picture pixelates, we don't notice this "cheating." Similarly, we don't notice the two kinds of signals coming from our eyes, because information from the "Where" system about motion, depth, and location is integrated with signals about color and object identity.

Once again, however, visual experience is fundamentally hermeneutic, because it entails both selection and combination. Inputs from the external world are filtered and differentiated according to the variable sensitivities of the receptors on the retina (rods and cones) and of the pathways transporting them (large and small ganglion cells). These separate, distinctive signals are then structured into coherent patterns by the reciprocal interactions among visual systems within the cortex. The brain makes it possible for us to see by combining parts into meaningful wholes that in turn give meaning to the parts.

Color is a particularly striking example of the brain's hermeneutic activity. It is a fundamental principle of neuroscience that "there is no such thing as color in the physical world; there is simply a spectrum of visible wavelengths that are reflected by objects around us."[18] The fact that our eyes respond to a particular, very limited set of wavelengths (whereas other animals have different ranges of response) is a result of evolution that is both arbitrary, in the sense that it could have been otherwise, and pragmatically necessary, in the sense that the construction of color has developed as it has because it has been useful in a variety of ways. As Zeki explains, "Colour is the result of the operation that the brain undertakes on the information that it receives; it is, in a real sense, a property of the brain and not of the world outside, even if dependent upon the physical reality in that world" (*Inner Vision* 185–86). Although vision produces color by processing wavelengths emanating from the external world, the perception of color is not simply the result of a one-to-one correspondence between ranges of wavelength and photoreceptor response. Rather, complicated calculations of ratios are necessary that are both internal and external. Internal to the visual system, as Livingstone explains, "our color perception depends on the ratios between the three classes of cones, not on the precise wavelength composition of the light" (97). The color we perceive is a complex construction in which the responses of different cones are aligned and opposed, and the well-known ability of the three "primary colors" to combine into all the shades of the rainbow is a consequence of these internal ratios (see Livingstone 26–35).

Similarly, we perceive color constancy despite external shifts in the illumination of an object because of a different but related ratio calculation. According to Zeki, color constancy is a product of the brain's ability "to discard all the variations in the wavelength-energy composition of the light reflected from a surface and assign a constant color to it" by constructing a ratio of the waveband reflected from the surface (A) and adjacent areas (B), a ratio (A/B) "which will therefore always remain the same in all illumination conditions" even if the absolute values of A and B change; consequently, "the construction of a color . . . is independent of the precise wavelength composition of the light reflected from it" but is "a matter of comparison between one surface and surrounding surfaces." Perhaps surprisingly reminiscent of Saussure's well-known analysis of the diacritical construction of the linguistic sign, Zeki describes color as "a visual language" based on "comparison . . . without reference to absolute values" ("Neurology of Ambiguity" 179, 180). As Saussure famously declares, "In language there are only differences *without positive terms*."[19] Color is not entirely arbitrary, of course, in the manner of the linguistic sign, because it is at least in part a response to wavelengths that impinge upon the retina, but those wavelengths do not absolutely and unequivocally determine by themselves the color we perceive.

Color is not an absolute, then, but a hermeneutic construct in at least two ways. First, like language, it is a diacritical and synthetic production of meaning based on selection and combination (cones responding differentially to wavelengths, and the brain synthesizing these responses from different alignments of cones to create various shades of color). Second, color constancy uses the unchanging ratios between surfaces to create the useful fiction of stability so that we do not get hopelessly lost in the Heraclitean flux, suppressing variations and fluctuations in wavelength illumination that would otherwise make objects seem to be ever shifting their values.

Constancy has its uses, then, but it can also be a trap. How does the brain recognize when differences, irregularities, and discontinuities matter and should not be suppressed? How does it decide that it needs to change the pattern it has constructed? These questions suggest, again, the problem of vicious hermeneutic circularity—or in biological terms, the dilemma of how an organism negotiates the competing values of stability and flexibility. Zeki makes the strong case for constancy: "The brain is . . . only interested in the constant, non-changing, permanent and characteristic properties of objects and surfaces in the external world, those characteristics which enable it to categorise objects" even if, or precisely because, "the information reaching it from that external world is never constant" (*Inner Vision* 5). The regularities we recognize in the irregular details reaching us through our various sensory systems enable the hermeneutic construction of patterns (or "categories") that create meaningful relations between parts and wholes, and these gestalts are useful navigational tools. As the cognitive literary theorist Ellen Spolsky notes, however, "our brains . . . are what biologists call open systems, meaning that they feed on stuff from outside themselves," and "that outside stuff doesn't stay still for long," so that "the maps our brains make are never finally or completely calibrated. . . . There are, thus, always discrepancies or gaps, and the brain is always working to catch up with itself."[20] The construction of patterns is useful for the very reason that they help us deal with the ever-changing flux—what William James memorably called the "great blooming, buzzing confusion" of the world—but the world's discontinuities and irregularities defy complete, comprehensive synthesis and can make those ordering fictions ineffective and even dangerous if they rigidify.[21]

The brain's quest for constancy is inherently contradictory. On the one hand, the value of constancy depends on its effectiveness in managing (and not simply ignoring) the changeableness of its inputs, and so the brain must remain open in order to respond to these irregularities. On the other hand, simply remaining open to novelty is ineffective because the brain can only identify a significant change against a background of constancy. A discontinuity can emerge only because it interrupts continuity. The paradox of the hermeneutic circle's need for both pattern and discontinuity and of the brain's need for both stability and flexibility is that constancy and openness to change require and depend on each other even as they oppose and exclude each other.

The brain has evolved various ways of handling this conflicting imperative, although the question of how it responds to novelty is still a somewhat murky area of neuroscience.[22] The brain's solutions begin at the neuronal level and go all the way up to the overall organization of the brain and its capacity to "play." At the cellular level, the neurons in the retina, which receive and transmit visual signals, have a "center/surround" structure (see fig. 3.3), which makes them especially sensitive to difference and change. As Livingstone explains,

> Center/surround organization makes cells at these early stages of the visual system sensitive to discontinuities in the pattern of light falling on the retina, rather than to the absolute level of light. Because of center/surround organization, neurons respond best to sharp changes, rather than to gradual shifts in luminance. The visual system is wired up in this way so that it can ignore gradual changes in light and the overall level of the illuminant, which are usually not biologically important. (54)

These cells respond in opposite ways to stimulation in the center and in the surround; light that excites the center, for example, inhibits the surround. Consequently, in the case of uniform illumination across the cell, the positive and negative reactions of the two opposing areas cancel each other out, and no response is registered. By contrast, a sharp ray of light beamed exclusively onto the center will cause excitement without counteracting inhibition. Sim-

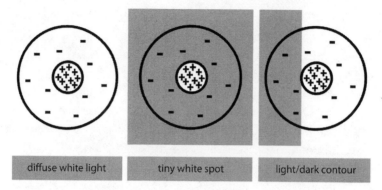

Figure 3.3. Responses of center/surround cells to diffuse light, spotlight, and edges. An illustration of why center/surround cells respond more to focused light or to a sharp edge than to diffuse, evenly distributed light. The activation of the center is canceled out by the inhibition of the surround with diffuse light, whereas with focused light or an edge the different rates of activation and inhibition in the two areas (the pluses and minuses aren't equal) register a response. From *Vision and Art*, by Margaret Livingstone, © 2002. Published by Abrams, an imprint of Harry N. Abrams, Inc., New York. All rights reserved. Reproduced by permission.

ilarly, a line that crosses the cell and stimulates differently parts of the center and surround will set off an unequal array of reactions (a differential mix of positives and negatives that do not balance out), and this too will register as a meaningful discontinuity.

There are various advantages to the center/surround organization (sometimes called *opponency*). As Livingstone notes, "It is much more efficient to encode only those parts of the image where there are changes or discontinuities than to encode the entire image," and "the most information in an image is in its discontinuities" (54–55). Again, the engineering behind high-definition television follows a similar principle, transmitting signals only about changes in the projected image rather than a much more voluminous, continuous stream of information about every single pixel on the screen.

How to process these signals about discontinuities and differences—and how to distinguish irregularities that matter from the normal flux and flow—is still a question, however, and here the asynchronous, imperfectly homogeneous interaction of various processing systems in the brain becomes important. The brain's normal state is an unstable balancing act between different, partially autonomous operations. Biologically considered, this is not particularly unusual, as Francisco Varela points out: "Biological systems demonstrate *instability* as the basis of *normal* functioning."[23] Even within the visual system, for example, recall that motion, color, orientation, location, and facial and object recognition are processed in different places and at different speeds. The integration between these areas is consequently always imperfect and incomplete and open to revision. Play between harmony and dissonance is not unusual, then, but is an ongoing, typical characteristic of the reciprocal relations among the brain's related, connected, but distinct and always somewhat disjunctive systems of information processing.

This incomplete integration makes it possible for the brain to entertain conflicting interpretations of the inputs it receives. The neurobiology of hermeneutic conflict begins here. As Zeki explains, "There are different levels of 'ambiguity' dictated by neurological necessity and built into the physiology of the brain" (*Splendors* 63):

> If the brain did not have the capacity to project more than one interpretation onto the stimulus, it may find itself in a dangerous situation. A good example is that of a smile on the face of someone one may fancy. If one were to give only one interpretation to that smile—a desire for greater intimacy—one may soon end up in trouble. Better for the brain to entertain several possibilities and thus protect itself. (62)

Discrepancies between different visual processing areas, or between different systems, such as vision or hearing (as in the visual and aural pathways to reading and grapheme-phoneme recognition), make possible an experimental to-and-fro between different configurations of meaning, and the fluidity of these interactions works against the tendency of established patterns to rigidify.

Such play between alternative interpretations is most in evidence in the case of ambiguous figures like the rabbit/duck and urn/two faces gestalts, and this is perhaps why the perception of ambiguity has been of central importance to neuroscience, as it has also been to aesthetics. Zeki offers what he calls "a neurological definition of ambiguity": "It is not vagueness or uncertainty, but rather certainty, the certainty of different scenarios, each of which has equal validity with the others" (*Splendors* 88). This definition is consistent with E. H. Gombrich's well-known argument that it is impossible to hold both constructions of an ambiguous figure in one's mind simultaneously. As Gombrich explains, "To see the shape apart from its interpretation . . . is not really possible. True, we can switch from one reading to another with increasing rapidity; we will also 'remember' the rabbit while we see the duck, but the more closely we watch ourselves, the more certainly we will discover that we cannot experience alternative readings at the same time."[24]

The literary theorist W. J. T. Mitchell notes Ludwig Wittgenstein's dissent—his argument that it is indeed possible to see such a figure as neither a duck nor a rabbit but as a "Duck-Rabbit," as a "metapicture" (Mitchell's term) that, as an exemplary instance of ambiguity, can be recognized as a particular kind of image. Mitchell observes, however, that Wittgenstein's point is the same as Gombrich's, that is, "the impossibility of getting outside the picture, except into another picture."[25] In neurological terms, each of these three configurations (duck, rabbit, or duck-rabbit) would entail a somewhat different cortical synthesis, creating a different assembly of neurons, each temporarily holding the others at bay but capable of yielding its hold on attention as one assembly dissolves and another emerges and becomes dominant. That is why they are felt to be mutually exclusive perceptions and different states of consciousness. Neuroscientists call such figures multi-stable images because they can stabilize in a variety of incommensurable configurations. The capacity of such images to switch between stable states—the instability of their stability—is what makes them ambiguous and intriguing.

Not all ambiguity is equally unstable, however, and the varying degrees of instability of different multi-stable images provide important clues about how the switching between them takes place—and, therefore, about how the brain negotiates the competing claims of constancy and flexibility. Zeki suggests that "there are graded steps, not only from non-ambiguous to ambiguous stimuli, but also in the number of areas or distinct cortical sites that may be involved during the perception of what are called ambiguous figures" (*Splendors* 91). Among the most automatic (and therefore least ambiguous) of these figures would be the so-called Kanizsa triangle (see fig. 3.4), which, as Zeki notes, "the brain tries to make sense of . . . by 'finishing it off' in the most plausible way, and interprets the pattern of luminances . . . as a triangle" ("Neurology of Ambiguity" 181). This figure is still ambiguous, however, precisely because we can perceive the invisible triangle as something that is paradoxically both "there" and "not-there" (a product of our "finishing off"). Otherwise it would be just another triangle, with nothing intriguing or unstable about it.

Based on evidence from brain-imaging experiments, Zeki suggests that the near-automatic perception of the Kanizsa triangle "is probably dictated by

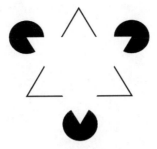

Figure 3.4. The Kanizsa triangle. Drawing by author.

the physiology of orientation selective cells . . . in areas V2 and V3," which
"are capable of responding to virtual lines." He notes, however, that this
figure also "activates area LOC, . . . a processing-perceptual centre for objects," which typically "works in collaboration with areas V2 and V3, . . . with
which it is presumably reciprocally connected" ("Neurology of Ambiguity"
181–82). The paradoxical quality of this figure is no doubt attributable to
slight disjunctions in the interactions between these areas. The quasi-automatic production of the invisible triangle may indeed be evidence of how
orientation-selective cells combine with object-recognition neurons in LOC
and reciprocally produce an image almost without our noticing anything unusual. But the discrepancies between them—the absence of the object that
the LOC area otherwise recognizes as three lines intersecting at three corners, for example, but that the orientation–selective cells want to produce
even though the evidence is missing—are probably what make the image
shimmer and refuse to settle down in the way it would if the missing figure
were filled in (see fig. 3.5). These very small but nevertheless distinct anomalies between cortical areas allow for the brain to "play" with interactions that
would otherwise quickly stabilize into a single, constant construct. Instead,
this play makes the figure shimmer unsteadily as a paradoxical absent presence. This shimmering in turn lays bare the tendency of the brain to "finish off" and fill in the gaps in incomplete evidence, a process that remains
invisible as long as it functions seamlessly, without the sort of dissonance in
evidence here.

A slightly more ambiguous bi-stable image opens up further room for intracortical play. The so-called Necker cube (sometimes also referred to as the
Kanizsa cube) tends to flip-flop as the lower and upper squares jump back

Figure 3.5. The Kanizsa triangle filled in. Drawing by
author.

and forth, changing recessional planes, moving closer or farther away (see
fig. 3.6). Zeki comments:

> It is difficult to tell whether this interpretational flip-flop is due to any "top-
> down" influences, that is, to activity of areas beyond the ones that register and
> combine the oriented lines into particular groupings. Imaging experiments
> show that every time the interpretation shifts from one plane to another, activ-
> ity in area V3 increases. But they also show that there is an activation of the
> fronto-parietal cortex. The interpretation of the latter result is not straightfor-
> ward; it might be due to sudden surges and shifts of attention, since the fronto-
> parietal cortex is known to be involved in attentional states. Or it might be
> due to some top-down influence that dictates that a shift in perception should
> occur. Either way, the result is substantially different from that obtained with
> viewing colors, when there is no activation of the fronto-parietal cortex.
> (*Splendors* 78–79)

These scanning data have at least two interesting implications. First, in contrast to the near-automatic perception of the Kanisza triangle, which involves
only adjacent areas of the rear visual cortex, the interactions recorded here
between the V3 area (neurons selectively sensitive to orientation) and the
frontoparietal cortex demonstrate the kind of long-distance connection that
makes possible reciprocal interactions between widely separated areas of the
brain.

Second, although the role of the frontoparietal cortex in attention is
widely documented, the difficulty of specifying exactly what it is doing here
is very possibly a symptom of a bottom-up and top-down interaction.[26] Ei-

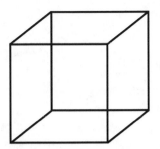

Figure 3.6. The Necker (or
Kanizsa) cube. Drawing by
author.

ther the frontoparietal cortex is stimulated in a bottom-up manner because
the image refuses to settle down and thereby calls attention to itself, or the
brain's mechanisms for directing attention try to guide perception in a top-
down manner, compelling one or the other of the possible gestalts to emerge
by emphasizing one or the other recessional plane. Or, most likely, both processes are occurring in a reciprocal, back-and-forth interaction as the images
flip-flop.

Even more complicated responses, involving a greater number of areas of
the brain, are set in motion by more complex multi-stable images, including
works of art. According to Zeki, "There is a continuum in the operations of
the brain . . . from conditions where the brain has no option in its interpretation of the signals that it receives, as in color vision, to those in which there
are two equally plausible interpretations and, finally, to those in which there
are many interpretations" (*Splendors* 96–97). These last cases are, understandably, more difficult to analyze with precise brain-imaging experiments.
Nevertheless, Zeki seems on safe ground when he suggests that "the true ambiguity that is often a characteristic of great art" involves "other areas, beyond
the essential nodes," where the visual cortex automatically processes stable
images (86):

> One can therefore conjecture that there are graded steps, not only from non-
> ambiguous to ambiguous stimuli, but also in the number of areas or distinct
> cortical sites that may be involved during the perception of what are called
> ambiguous figures. At the highest levels, as evidenced by the capacity to give
> multiple, equally valid, interpretations to a work of art, the ambiguous state
> may involve several distinct areas that are able to bring their influence. . . .

Here memory, experience, learning and much else besides can influence what is perceived at any given moment. This almost certainly involves a "top-down" influence from diverse sources, not just the frontal lobes. (*Splendors* 91)

With semantically complex figures, where more associations would be triggered than by geometric shapes, the hippocampus and other memory sites would no doubt be activated as past experiences and learned conventions are brought to bear in interpreting new configurations of meaning, as would the frontal lobes, which, as Elkhonon Goldberg observes, "are critical in a free-choice situation, when it is up to the subject to interpret an ambiguous situation."[27]

Although the experimental data are somewhat rudimentary, Goldberg reports that novelty (such as one typically encounters with works of art) sets in motion an interaction between areas in the left hemisphere, which specialize in routine operations, and areas in the front right hemisphere, which are activated by open-ended, unpredictable experimentation. He cautions that "in reality each cerebral hemisphere is involved in all the cognitive processes, but their relative degree of involvement varies according to the novelty-routinization principle," which he explains with a musical analogy: "It appears that the cerebral orchestra is divided into two groups of players. Those sitting on the right side of the aisle are quicker at basic mastery of the new repertoire, but in the long run, with due practice, those on the left side come closer to perfection" (79).

This interaction of cortical areas for routine and novelty is consistent with another well-known formulation of the hermeneutic circle, namely, that one can only understand the unfamiliar by grafting it onto what is already familiar, even if this causes the unfamiliar to transform the familiar. The right side of the orchestra may be more open to novel phenomena and better able to improvise than the left side, but an unfamiliar state of affairs can only be assimilated by altering the patterns that the left side has mastered. If these techniques aren't sufficient, the right side's improvisations must still take off from and extend (and thereby transform) what the other side commands. How the brain can experiment depends on the repertoire it already has at its disposal—the typical neuronal assemblies that it forms in response to habitually encountered stimuli, accustomed cortical patterns, which must be altered to account for anomalies. The paradox of the assimilation of novelty is that the familiar configurations, which cannot account for it, are the equip-

ment the brain must nevertheless use to make sense of new, strange, unprecedented phenomena. Both sides of the brain interact in the cycle of novelty
and routinization in a mutually formative, reciprocally interdependent manner because of this circular relation between the familiar and the unfamiliar.

Because what is familiar to two different brains will vary according to their
past interpretive experiences, they will necessarily experiment in dissimilar
ways with new hermeneutic challenges, and this is another potential source
of interpretive conflict. Our brains differ not only in what they know but also
in how they know it—their habitual repertoire of responses, reinforced by
Hebbian learning (neurons wiring together as they fire together). As the neuroscientist György Buzsáki points out, "The fine connectivity in the brain is
flexible and perpetually changing," and "no two brains have identical connectivity, in contrast to the rigid, blueprint-determined wiring of machines."[28]
These neuronal differences, established by past patterns of cortical activity,
are at work, for example, when a critic accustomed to reading literature for
its social and political relevance seeks different configurations of meaning
in a text than does a formalist, who is inclined to attend to how a novel or a
poem employs, revises, or breaks established linguistic conventions. Interpreters with opposing hermeneutic allegiances have differently wired brains
as a consequence of their histories as readers, and they will differ accordingly
not only in what they recognize as familiar when they encounter a new text
but also in how they experiment with building coherent part-whole relations
to make sense of the unfamiliar.

These interpretive disagreements are further evidence of the brain's contradictory combination of constancy and flexibility—its tendency to routin-
ize repeated operations and its capacity to reorganize itself in response to
novel stimuli. Confronted with novelty or anomaly, an interpreter does not
erase his or her brain's habitual patterns of response and start over again
from scratch but, rather, revises and extends the familiar to accommodate
the unfamiliar. The brain's sense of the world's constancies reinforces itself
over time (neurons firing and wiring together, over and over again), and two
different brains wired in different ways will consequently find that the divergences in their patterns of understanding will solidify and become more
firmly demarcated with repeated experience. So it is not surprising that interpreters who prefer different ways of reading often find themselves dividing
into opposing camps that may see the textual world in radically incommensurable ways.[29] But these differences can only arise in the first place because

the neuronal assemblies that the brain is capable of creating are flexible, at least to some degree. Some cortical responses are automatic and inherited, like color perception (although even here there is room for variation in how different brains respond to the same wavelengths of light).[30] But the different configurations of meaning that ambiguous, multi-stable images invoke are intriguing to neuroscience because they demonstrate the brain's capacity to reshape its internal connections and construct opposing patterns of meaning. This flexibility makes possible the divergences in wiring between two brains that can lead readers to see different patterns of meaning in the same text or that can cause one to see a rabbit where another finds a duck.

In all of these ways, then, the irreconcilable interpretive conflicts that characterize the humanities can be seen to have a basis in the neurobiology of the brain. This is not to say that neuroscience can resolve these disagreements or help us choose between opposing critical methods. But these disputes would not arise in the first place if they were incompatible with the way the brain works. The conflict of interpretations in the humanities, without the availability of appeal to a neutral judge, is a social, cultural manifestation of the competition between alternative patterns by which the brain negotiates its cognitive way through the world. For neuroscience, opposing interpretive constructs are both a demonstration of and a response to the conflict between the imperatives of stability and flexibility in cognition. Multiple alternative patterns for configuring inputs are advantageous because their competition keeps us flexible in responding to the ever-changing flux of the external world and gives us the possibility of not remaining trapped in one particular construct. But the fact that the brain cannot focus attention on two mutually exclusive configurations simultaneously, but must flip-flop between them, shows that cognition cannot take place by simply opening oneself up to the flux of phenomena. Alternative constructs make the brain flexible and open to change, but the conflicting demands of stability and flexibility can only be negotiated by shifting between patterns. There is no neutral third party, no central controller or man in the machine, to decide between them. The brain cannot evaluate competing patterns except with another pattern. Everything depends once again on the brain's ability to "play."

The brain's innate receptivity to multiple, conflicting interpretations is further demonstrated by its peculiar resistance to attempts to stabilize ambiguous figures. For example, if one introduces a sitting stick figure into the so-called staircase illusion in an effort to compel one reading of the reces-

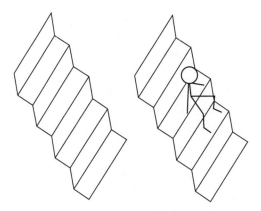

Figure 3.7. The staircase illusion. A demonstration of the persistence of ambiguity. Even when a seated stick figure is added, the recessional planes still "flip." Reproduced by permission from Semir Zeki, *Splendors and Miseries of the Brain: Love, Creativity, and the Quest for Human Happiness* (Malden, MA: Wiley-Blackwell, 2009).

sional planes occupied by the otherwise flip-flopping rectangles, one finds that the stairs nevertheless still tend to jump back and forth (see fig. 3.7). According to Zeki, this is typically the case with attempts to disambiguate ambiguous figures:

> Adding a number of features to the figure, to force the brain to interpret it in one way only, is never successful. The brain retains the options of interpreting it in two ways. This suggests that the brain does not have much choice in the multi-interpretations that its organization makes possible. . . . The ambiguous or unstable system of the brain is therefore highly stable in its instability. (*Splendors* 84–85)

Zeki interprets this phenomenon as evidence of the pragmatic value to the brain of keeping its options open: "Where one solution is not obviously better than the others, the only option is to allow of several interpretations, all of equal validity" ("Neurology of Ambiguity" 188). That, however, can also be paralyzing; hence the brain's attempt to forge a synthesis and establish a constant meaning, even as this integration is disturbed by residual dissonances.

This contradictory response makes good sense. If a predator seems to be approaching, one must act or die, but if the foe turns out to be a friend, re-serving the possibility of a change of course is useful too. Hence the apparent evolution of the brain's paradoxical response to this dilemma—holding on to only one reading at a time but keeping ready the capacity to switch frame-works, and not simply suppressing dissonant signals, even when the evidence for a particular reading (how the stairs fit under the sitting figure) seems over-whelming. Such resistance to a single, stable construal would be inefficient and counterproductive except for the pragmatic value of provisionality, the usefulness of resisting the trap of vicious hermeneutic circularity. Against the danger of getting stuck in a rigid and inflexible imposition of constancy, the disjunctions in cortical syntheses of conflicting signals hold open the pos-sibility of change.

These examples from the visual system illustrate well how the brain builds consistency, but they may make the hermeneutics of brain functioning seem predominantly cognitive. As Antonio Damasio has persuasively argued, however, emotions also play an important role in the brain's ability to as-sess situations and make judgments, and these processes similarly entail re-ciprocal part-whole configurations. As Damasio points out, "Our brains can often decide well, in seconds, or minutes, depending on the time frame we set as appropriate for the goal we want to achieve," and this ability requires "more than just pure reason." He proposes that "somatic markers"—feelings connected to particular body states—"increase the accuracy and efficiency of the decision process" by focusing attention and highlighting likely dan-gers or positive outcomes: "When a negative somatic marker is juxtaposed to a particular future outcome the combination functions as an alarm bell. When a positive somatic marker is juxtaposed instead, it becomes a beacon of incentive."[31] Patients with damage to emotional centers in the prefrontal cortex lack the ability to make good judgments about likely costs and benefits because somatic markers do not alert them to the shape of the situation they face. Damasio cites the well-known case of the nineteenth-century railroad worker Phineas Gage, whose life fell apart when an iron rod penetrated the emotional centers of his brain and he lost the capacity to make reasonable moral and social judgments about the consequences of his actions. Damasio also notes gambling experiments in which patients with damaged emotional regions showed themselves less able than a control group to learn from ex-perience and identify and avoid particular high-risk piles of cards. In both cases an emotional deficit impaired the ability to anticipate and recognize meaningful patterns.

Reason and feeling are not necessarily opposed, then, as the Cartesian model assumes, because bodily based emotions can have an important hermeneutic function. The value of somatic markers to judgment is their intuitive suggestion of part-whole relations to guide interpretation. Emotional, embodied intuitions give judgment an anticipatory sense of the patterns into which past experience suggests parts are likely to form, even as working through the parts is necessary to test and confirm (or revise) the expectations that feelings project. Emotions and cognition are closely linked because both operate by analogous configurative processes. Rather than antagonistic or mutually exclusive faculties, reason and feeling are linked by their participation in the hermeneutic circle as embodied processes of cognition.

The brain's capacity to alternate between conflicting readings is a reflection of its normal functioning—a key element of its stable instability—as syntheses come and go in response to ever-changing stimuli. It is not as if a man in the machine watches images projected by a camera and then assigns a value to them, as would be the case if the brain first ascertained the "meaning" of the world before it and then assigned a "significance" to this perception from a variety of competing possibilities. That is a distinction proposed by the influential monist E. D. Hirsch Jr., who claims that interpretation must first ascertain the determinate meaning of a state of affairs, to which varying and opposing significances may then be assigned as this meaning is set against different contexts or is deployed for competing purposes.[32] The neurological evidence speaks powerfully against this argument. As Zeki points out, "There is not a separate site specialized for perceiving, as opposed to processing" ("Neurology of Ambiguity" 179), as Hirsch's distinction between meaning and significance implies. Perceiving and processing are not separate functions but are integral aspects of the same operation of making sense of the world.

The potential for multiple, conflicting interpretations is there from start to finish as the reciprocal interactions between cortical regions produce provisional configurations of constancy. The brain constructs meaning by engaging in reciprocal, mutually formative play between cortical regions, and as in the brain's response to multi-stable figures, this play may alternate between conflicting modes of part-whole configuration. The possibility of contradictory, mutually exclusive syntheses within the brain can lead to disagreements between different brains (or in the same brain at different moments in its history) about the most effective, useful relation between parts and wholes. The resistance of multi-stable images to resolution suggests that the brain is hard wired for conflicting readings. The possibility of constructing opposing, mutually exclusive configurations of meaning is not an aberration, but is a reflection of processes in the brain that have evolved to negotiate the competing claims of constancy and flexibility. The interpretive conflict that is characteristic of the humanities has deep roots in the neurobiology of mental functioning.

Given the configurative hermeneutic processes that characterize the visual system, it is not surprising that similarly circular pattern recognition is evident in the construal of letters and words. For example, in a well-known experiment by Gerald Reichler, literate adults are presented the same letter (*D* or *T*) either alone on a screen or as part of a word (*HEAD* or *HEAT*), and it turns out that the accuracy of recognition is much worse when the letter is seen in isolation than when it is seen in the context provided by the word.[33] The gestalt of the word *HEAD* or *HEAT* is what matters, not the linear sequence of letters, because the string *HEA* is the same in both cases, and yet adding the seemingly uninformative, redundant *HEA* makes the *D* and *T* more recognizable because of the word pattern *HEAD* versus *HEAT*. The same result even occurs with configurations of letters that look like word shapes typical of the reader's language (*GERD* or *GERT*) or strings of consonants that are similar to actual words (*SPRD* or *SPRT*), but it no longer happens with random letters or strings that don't recall word shapes (*GQSD* or *GQST*).[34] This would not be the case if word comprehension were sequential, a linear addition of letter to letter, as in a scanning machine. The hermeneutic circle is at work even in visual letter and word recognition because this turns out to be a process of configuring part-whole relations.

Brain-imaging experiments confirm that the visual word form area is characterized by configurative processes. Dehaene reports that imaging experiments in his laboratory have demonstrated that the letterbox area "does not merely respond passively and innately to anything that merely resembles a letter or a word":

> Strings of letters do not always stimulate it equally well—spelling in the language familiar to the reader must be respected. For example, this region re-

sponds far better to strings that form an existing or plausible word, such as 'CABINET' or 'PILAVER,' than to strings that violate spelling rules, like a string of consonants such as 'CQBPRGT' [see fig. 3.8]. It also prefers customary letter combinations, like 'WH' or 'ING,' to rare or impossible ones such as 'HW' or 'QNF.' Even valid letter strings can fail to activate the letterbox area if the person who is scanned hasn't learned to read them—thus Hebrew characters provoke strong occipito-temporal activation in Hebrew readers, but not in English readers. (95)

These results lead Dehaene to speculate about the possible existence of "bi-gram neurons," which recognize "ordered pairs of letters" (154). Their existence is, he acknowledges, merely "an educated guess" at this point, but he proposes it to explain "similarity effects" such as our tendency to automatically correct certain letter reversals, so that "we experience little difficulty in raeding etnire sneetnecs in wihch the ltteers of eervy wrod hvae been miexd up, ecxpet for the frsit and the lsat ltteers" (154, 156). This, of course, is also why it is so easy to miss typographical errors when proofreading. These phenomena would be inexplicable if comprehension were additive and linear rather than a configurative construction in which the overarching pattern guides the construal of its parts.

Another example of the circular interdependence of letter and word recognition is provided by ambiguous handwriting. For example, we can decipher the script honey bccs' sweet nectar, even though the letters *e* and *c* have the same shape in bee and nectar (see Dehaene 160). Once again expectations provided by the context in which the parts are formed allow the interpretation of words that a linear scanning machine would find incomprehensible. In neuronal terms, "the constant interplay of bottom-up and top-down connections" in the brain allows this to happen, according to Dehaene, because "populations of neurons . . . constantly send messages in all directions, thus passing on the fragmentary data at their disposal to each other until the whole group converges toward an agreement" (160). The brain makes sense of the world by forming neuronal assemblies, and this is not a linear, additive process that follows billiard-ball causality. Patterns are created by reciprocal connections between different neurons in different parts of the brain, which exchange signals back and forth, the semantic recognition of a word pattern joining bee, nectar, and honey interacting with letter-form identification processes and, here, overriding anomalies that don't fit the expected configu-

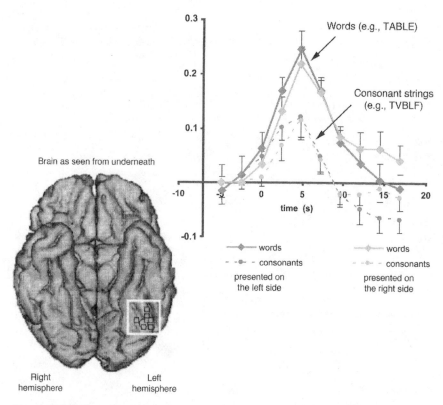

Figure 3.8. The letterbox's response to words compared with its response to letter strings. The letterbox is activated more by real words (e.g., TABLE) than by consonant strings that violate the rules of the experimental subject's language. Adapted by permission from Laurent Cohen et al., "Language-Specific Tuning of the Visual Cortex? Functional Properties of the Visual Word Form Area," *Brain* 125.5 (2002): 1054–69.

ration. The bottom-up and top-down interplay of brain areas through this back-and forth signaling makes possible the reciprocal alignment of parts and wholes necessary for comprehension, translating into meaningful configurations spelling errors that would baffle a scanning machine. Here as elsewhere, the to-and-fro play through which neuronal assemblies form is the neurobiological basis of the hermeneutic circle.

These reciprocal interactions are also evident in phoneme-grapheme translation. Ample experimental evidence shows that the ability to recognize phonemes (as opposed to nonlinguistic sounds) is enhanced by literacy. The interdependence of the recognition of spoken and of written signs is powerful evidence of the reciprocal interaction of the phonemic and graphic pathways to language, and this interaction is only hermeneutically useful because each system provides the other with patterns that help make sense of its constituent parts. A classic experiment by the linguist José Morais demonstrated, for example, that "illiterate adults were unable to delete or add a phoneme at the beginning of a non-word, while adults from the same environment [a rural Portuguese village] who learned to read in youth or as adults had little difficulty."[35] In a later, follow-up experiment, Morais showed that illiterate adults were less able than readers to detect and delete a phoneme from a word they knew or to recognize a rhyme.[36] Commenting on these experiments, Dehaene infers that illiterate adults or practitioners of pictographic, nonphonemic scripts would not understand the joke in spoonerisms like "Our Lord is a shoving leopard" (instead of "loving shepherd") because they could not make the graphic-phonemic substitutions such plays with word forms and word sounds entail. He concludes: "The deep effects of phonemic awareness prove how profoundly the acquisition of the alphabetic code changes our brains" (202).

These changes make possible a reciprocal exchange of information between the visual and auditory systems that has consequences for the patterns to which each responds. Graphic gestalts allow auditory forms (phonemes) to be recognized that are otherwise invisible in the flux of phenomena, even as these auditory structures help make sense of written words. This reciprocal interaction would not occur if these two domains were linear and additive processes in which each adjacent part combined sequentially with the next to construct a summative meaning. Only because graphic forms give the brain an anticipatory sense of the patterns into which phonemes will form (as elements in written words) can literacy provide resources for understanding the sound patterns of language that are invisible to adults who can speak but not read. This point is also demonstrated by the well-known phenomenon of *oronyms* (e.g., "The good can decay many ways" and "The good candy came anyways"), in which the same sounds can be construed to have different meanings depending on the graphic word forms they are seen to take. Oronyms pose a difficulty for computer voice-recognition programs, Hubert Dreyfus points out, because linear scanning processes have a hard time dealing with recursive phenomena in which the recognition of parts depends on their overall configuration, as is the case here, where "the same physical constellation of sound waves is heard as quite different phonemes, depending on the expected meaning."[37]

The hermeneutic circle is not a phantom of philosophy and literary theory, then, but is powerfully confirmed by empirical psychology and neuroscience. Nevertheless, the seeming mystery of processes in which comprehension of the whole precedes and directs the construal of the parts makes some linguists wrongly wary of this circular (or spiral) metaphor even if they acknowledge that recursivity and reciprocity are essential to how language works. For example, although the popular linguist Stephen Pinker warns that linear "parsing" models and "word-chain devices" cannot account for the "combinatorial" effects of language, he nevertheless insists that language "is compatible in principle with the billiard-ball causality of the physical universe, not just mysticism dressed up in a biological metaphor."[38]

Hermeneutic circularity is not billiard-ball causality, but neither is it mysticism. Pinker himself notes that "a property of language" is "its use of 'long-distance dependencies' between an early word and a later one," and he acknowledges that "word-chain devices cannot handle these dependencies" (89). Some of these dependencies are governed by syntax (as with the German verbs that predictably turn up at the end of dependent clauses), and others are a matter of semantics and context that require complex acts of interpretation. Colliding "billiard balls" cannot account for such relations, either syntactically or semantically, because sequential explanations cannot capture the circular, reciprocal part-whole interaction that the construal of long-distance dependencies requires—the to-and-fro movement through which we recognize, for example, that *too* means "also" and not "two" in this series of interdependent meaning units: "You too can go to the movies. I'll give you some money so that you can accompany your friend John. I don't want you to feel left out," as opposed to "You two can go to the movies. After studying so hard for your mid-terms, both of you have earned some fun."

Another linear metaphor that Pinker employs, a branching "tree," is a little better, but not much. Pinker proposes that "phonemes are not assembled into words as one dimensional left-to-right strings. Like words and phrases, they are grouped into units, which are then grouped into bigger units, and so on, defining a tree" (169). The figure of a tree is linear and sequential, however, and the grouping implied by such a metaphor is insufficiently interactive to account for the to-and-fro configuration of meaning in a long-term dependency. A tree does not curve back on itself, with later branches potentially changing in retrospect the meaning of earlier ones, as in the way the second

Figure 3.9. Branching and recursive trees. A fanciful depiction by the anthropologist Alfred L. Kroeber of how the image of a "branching tree" would need to be modified in order to represent recursive, reciprocal interactions. Reproduced by permission from A. L. Kroeber, *Anthropology* (New York: Harcourt, Brace, 1948), 260.

sentence in these examples redefines the meaning of *too* or *two* in the first sentence.

The anthropologist Alfred Kroeber offers a fanciful depiction of how the branching-tree figure would need to be revised in order to represent recursive phenomena, with various of the branches turning back and growing into one another, rather than every branch subdividing sequentially into a series of forks (see fig. 3.9). Indeed, a more rigorous application of biological knowledge to the tree metaphor might introduce the possibility of such interactions (leaves and roots viewed, for example, as interdependent elements in the system of photosynthesis, through which the tree breathes, transforms nutrients, and grows, and sometimes separate branches do merge again). That is not, however, the logic typically implied by the convention of graphing a tree as a series of forks. That kind of branching is consistent with the billiard-ball metaphor, inasmuch as causality is depicted as a linear sequence, branching and then branching again—but not circling back to affect processes earlier in the chain of divisions.

Pinker's recursive explanation of nonlinear combinatorial linguistic processes is at odds with his metaphysical allegiance to Newtonian models of causality, and this conflict is evident in the confusion of his metaphors. Phonemes, graphemes, and other constituents of language are indeed "grouped into bigger units," which are then combined into larger units, but this combinatorial action is not all one way or simply sequential or a matter of branching. It is recursive and reciprocal, moving forward and backward, in a to-and-fro manner, with our recognition of larger units also turning back and retrospectively affecting the meaning of smaller units, which in turn generate expectations about the larger patterns into which they will combine (sometimes changing groupings we had provisionally formed). That is how a long-range dependency works, and this to-and-fro reciprocity is what the hermeneutic circle is all about. We don't need mystical hocus-pocus to account for it, however, because the circular, recursive characteristics of interpretation are a manifestation of the noncentered, reciprocal interactions that are normal to the functioning of the brain.

Aesthetic experiences of harmony and dissonance play with the brain's recursivity and its contradictory need to create constancy and to preserve flexibility. How this happens, with what potential consequences, is suggested by phenomenological accounts of reading as a process of gap filling and consistency building. These descriptions of reading are fully consistent with neuroscientific explanations of the hermeneutic circle, and they suggest how the neurological processes I have analyzed are manifested in our interactions with literature.

For example, reading a literary work is similar to the visual system in its tendency to "finish off" incomplete figures, inasmuch as we read by "filling in the gaps left by the text" (as Wolfgang Iser argues), whether these are indeterminacies left unspecified by the perspectives in which characters, objects, and scenes are represented or tacit meanings suggested but not explicitly articulated, or connections between states of affairs left for the reader to discover. These are only some of the many and various "unwritten" implications of the "written text" that make up the "virtual dimension" that we constitute in our experience of reading.[39] Gap filling encourages the reader's immersion in the text's world and the building of illusions. A text's blanks may also create a space for abstract readerly reflections. As Zeki notes, "There are substantial advantages to . . . leaving things unfinished" in a work of art and allowing "the synthetic concepts of the brain" to complete what the artist starts (*Splendors* 55), but this is the case with all art, and not only the actually unfinished works that Zeki cites. According to Iser, all literary works are unfinished, inasmuch as they invite the reader's participation by leaving blanks and indeterminacies for us to fill in. The same is also true with music and the visual arts, if only because the patterns through which the audience will give meaning to them are never fully specified but are virtual gestalts that we must recognize and construct.

Some gaps are representational—the indeterminacies in specifying a character's features, for example, which readers may or may not explicitly fill in, and about which they may not realize they had tacit assumptions until they are surprised, and perhaps disappointed, by an illustration or a movie. ("That's not how this character looks!" was my reaction to seeing Nicole Kidman as Isabel Archer in Jane Campion's *Portrait of a Lady*, but John Malkovich completed my vision of the villainous Gilbert Osmond in ways that have deepened my appreciation of the novel.) Part of the disappointment, indeed, may simply be that an indeterminacy has been specified and is thus no longer available for the imagination to play with (which is why Henry James disliked the practice of illustration). But other blanks are the connections between parts of a text that any work leaves as the reader's share, even (or especially) abstract and nonrepresentational art forms. The absent spaces between juxtaposed images or perspectives are blanks that call attention to the consistency-building activity of reading (recall Ezra Pound's modernist classic "In a Station of the Metro," which both incites and frustrates our quest for connection: "The apparition of these faces in the crowd; / Petals on a wet, black bough").

Different works can be characterized by the particular kinds of blanks and indeterminacies they offer to the reader, even as one's sense of what these are will change over history as new interpretive practices equip readers with novel expectations about the patterns they will find in texts. Different readers will fill gaps and make links across a text's blank spaces in different ways, and the indeterminacies that are "there" for some readers may not exist for others. For example, whether a character has unconscious sexual desires, a matter of some urgency for a psychoanalytic critic, is not a gap that other readers may notice unless it is explicitly dramatized. Or to take a related case, T. S. Eliot finds that Hamlet's indecision lacks any "objective correlative," and this indeterminacy is in his view a deficiency in Shakespeare's art, but a Freudian reader will fill this gap with suppositions about Oedipal guilt. Other readers have found this very indeterminacy to be, not a flaw or a symptom of repression, but a provocation for existential reflection. All texts have a virtual dimension, but this will be different not only for different texts but also for different readers based on their histories of reading.

The brain's inherent tendency to forge syntheses is invoked by the gaps and indeterminacies in a work, and the harmonies that we can construct out of the suggestions held ready by the text invoke, reinforce, or extend our pattern-forming capacities. The creation of pleasing harmonies sets in motion the to-and-fro play whereby the brain creates meaning, and these interactions may confirm and reinforce neuronal patterns in a Hebbian manner (neurons firing and wiring together). Or new interactions made possible by these harmonies may reveal new ways of constructing relations between parts and wholes and establish new intracortical connections. According to Iser, however, "a text may either not go far enough [in leaving things for us to fill in] or may go too far, so we may say that boredom and overstrain form the boundaries beyond which the reader will leave the field of play" ("Reading Process" 275). These boundaries will of course vary for different readers (which is why some of my students love *Ulysses* and others hate it, and I draw the line at *Finnegans Wake*, although for other readers that's where the fun begins). All brains like to play (speaking metaphorically), but the kinds of play that different brains are ready for may vary widely depending on the interactions they are accustomed to (the constancy they are familiar with) and the openness of these habitual patterns to change and variation (their flexibility in the face of novelty).

Frustration is not always an unpleasant experience, however. Surprise is a crucial, typical, and not at all unenjoyable aesthetic experience. Surprise occurs in our experience of texts because reading is a process not only of building but also of breaking consistency. A pattern must be built to be broken, and the play of a text in encouraging and then disrupting the formation of part-whole configurations is a not unusual aesthetic experience even with works in which harmony is ultimately reestablished (if in a different form than first suggested). As the cognitive psychologist Richard Gerrig has shown, reading even a simple, nonliterary text entails the deployment of "schemas" and "scripts" through which part-whole configurations of meaning are constructed.[40] Welcoming but then challenging the expectations readers bring to texts based on their habitual patterns of consistency building is one of the ways literature typically plays with the brain.

As Iser, Jauss, and other phenomenological theorists point out, one function of the disruption of consistency that a surprising, perhaps frustrating aesthetic experience provides is to reveal the limits of such habits and expectations, to lay bare their workings in ways we may not have noticed so long as they proceeded smoothly, without interruption. Gadamer finds special value in "the experience of being pulled up short by the text," because such disruptions disclose and test the prejudices (*Vor-urteile*) without which we cannot understand but which threaten to rigidify.[41] Gadamer may exaggerate when he claims that "only through negative instances do we acquire new experiences" (356), but the power of negation is an important factor in the balancing act between constancy and flexibility. Aesthetic experiences of dissonance resist the brain's quest for constancy in a way that may seem confusing and even annoying (because synthesis is thwarted), but these disjunctive moments may also seem playfully provocative and even emancipatory (because flexibility is enhanced).

Experiences of dissonance can lead to new modes of consistency building. This is especially evident in the cognitive power of metaphors to create new meanings through disorienting semantic disjunctions that then reorient our capacity to make connections. The *interaction theory* of metaphor pioneered by the pragmatist philosopher Nelson Goodman and the hermeneutic phenomenologist Paul Ricoeur contends that metaphors bring about semantic innovations by violating and extending the established rules of a language.[42] A metaphor (e.g., the oft-cited example "man is a wolf") is created by an interaction between a word and a context in which it seems both strange and fitting. As Goodman memorably claims, "Where there is metaphor, there is conflict" (69). The dissonance occurs because the meanings conventionally associated with the anomalous term are incompatible with its setting, and this incongruity sets the reader hunting for an extension of meaning to restore consistency and, with it, sense. Ricoeur consequently describes metaphor as "a bringing-together of terms that first surprises" and "bewilders" before the reader "finally uncovers a relationship" to resolve "the paradox" (27).

The equal importance of disorientation and reorientation makes it an oversimplification to describe metaphor as a process of establishing resemblances. As Nietzsche powerfully declares, metaphor entails "das Gleichsetzen des Nicht-Gleichen," making "equal" the "not-equal."[43] The cognitive scientist Mark Turner's influential term for this process, *conceptual blending*, is consequently at best imprecise and at worst misleading. He defines *blending* as "the mental operation of combining two mental packets of meaning—two schematic frames of knowledge, or two scenarios, for example—... to create a third mental packet of meaning that has new, emergent meaning."[44] This description is fine as far as it goes, but the term *blend* is too homogeneous and harmonious to do justice to the friction, conflict, and negation involved in the interactions through which semantic innovation occurs. In such operations the effects of dissonance and disjunction in disrupting consistency are as important as the search they launch for new syntheses. The interaction that produces metaphoric meaning requires both disruption and new coherence, and the resulting synthesis remains incompletely "blended" to the extent that it retains traces of the anomalies that produced it, dissonances whose force dies out when the new metaphor becomes completely assimilated.

Only then, when the metaphor is dead, does it become a "blend." Stephen Pinker has similar reasons for calling language a "combinatorial system," and not a "blending system" like paint mixing or cooking, in which "the properties of the elements are lost in the average or mixture," as when red and white produce pink or the flavors of different ingredients merge in a stew (76). A dead metaphor is a blend in Pinker's sense because the qualities, or "flavors," of its constituent parts have been forgotten and no longer resonate. When writers make the mistake of using mixed metaphors, that is typically because the figures they use are "dead" and no longer resonate with the paradoxical, contradictory meanings through which semantic innovation occurs (although these emerge again when we notice the error). An innovative metaphor that brings "new, emergent meaning" to the world is not a blend. (For example, contrast "My lover has rosy lips" with this image from "The Love Song of J. Alfred Prufrock": "When the evening is spread out against the sky / Like a patient etherised upon a table.")

A successful, truly novel metaphor holds in tension difference and similarity, disjunction and coherence, anomaly and revelation. This is because metaphor functions by invoking and then frustrating the brain's quest for constancy, only then to make possible a new configuration of part and whole. Metaphor plays with the brain's contradictory need for both constancy and flexibility by first overturning patterns that may have become rigid and constraining but then, after these disruptions, promoting new modes of consistency building. Both of the brain's fundamental imperatives—its quest for stability and its openness to instability—are inherent in the paradoxical powers of metaphor.

The brain's response to novelty and ambiguity gives it the chance to learn about itself. The experience of construing novel, puzzling states of affairs, whether an ambiguous figure or a striking metaphor, can promote epistemological self-reflection because of the brain's reaction to disruptions to its routine patterns. For example, Zeki notes that with "fauvist paintings in which objects are dressed in 'un-natural' colors," the automatic construction of color constancy is interrupted, and imaging experiments then show that "the frontal lobes become activated, as if in trying to solve a puzzle" (*Splendors* 43). This activation of the frontal lobes is evidence of the brain's response to novelty—the free experimentation with hypotheses that Goldberg's novelty-routinization cycle entails as the cortex playfully improvises and searches for a new configuration of stability. Frontal-lobe activity is also stimulated when we become aware of a change in perception (see Zeki, *Splendors* 83). It is thus a signal of the brain's becoming cognizant of its own processes. Interruptions in consciousness may become an incitement to self-consciousness, because a blockage in an otherwise automatic perceptual process, like the construction of color, can produce awareness of two kinds, not only of the discontinuity that needs to be addressed but also of the puzzle-solving activity that responds to it.

In reading, similarly, blockages to our quest for consistency may offer us the opportunity to become aware of our typical habits of pattern making and gap filling, epistemological processes that we remain happily blind to so long as they function smoothly. As Iser notes, when our normal, quasi-automatic hermeneutic functioning is interrupted and we are compelled to experiment with new hypotheses about part-whole relations, "the need to decipher gives us the chance to formulate our own deciphering capacity" (*Implied Reader* 294).[45] This is an oft-observed effect of ambiguous figures, as W. J. T. Mitchell points out: "We might think of the multistable image as a device for educing self-knowledge, a kind of mirror for the beholder" (48). The challenge of the bewilderment that their anomalous effects provoke is not only to experiment with new patterns for building consistency and restoring constancy but also to become aware of cognitive processes that have been exposed by being blocked, processes that become visible with special clarity as the brain plays with alternative ways of interpreting these elusive, intriguingly unstable figures. The cognitive self-consciousness made possible by the playful instability of ambiguous, multi-stable figures is, to be sure, yet another reason why they have attracted so much attention from aestheticians as well as neuroscientists.

If, as is sometimes claimed, artists resemble neuroscientists inasmuch as their works offer us insight into the functioning of the brain, they typically conduct their neurological experiments by playing with ambiguity and hermeneutic multiplicity.[46] By blocking routine, automatic cognitive processes and instead promoting playful, self-conscious hypothesis testing, artists invite us to join these explorations. When ambiguous or paradoxical figures interrupt the hermeneutic circle, we readers may not exactly become neuroscientists—art, science, and reading are different, distinctive practices, after all, and each offers a different take on the world. But these disruptions give us the chance to reflect about what typically happens beneath our notice when we read, and in doing so we can analyze the workings of our brains.

CHAPTER 4

The Temporality of Reading and the Decentered Brain

The lived experience of time is intuitively obvious—until we begin to examine it, and then it can seem deeply paradoxical, even scandalous. As Augustine famously asked: "What, then, is time? I know well enough what it is, provided that nobody asks me; but if I am asked what it is, and try to explain, I am baffled."[1] The problem begins with the simple phenomenon that William James describes in an oft-quoted passage from *The Principles of Psychology*: "The practically cognized present is no knife-edge, but a saddle-back, with a certain breadth of its own on which we sit perched, and from which we look in two directions into time." What James calls "the specious present" is our lived experience of time as "a duration, with a bow and a stern, as it were—a rearward- and a forward-looking end."[2] This durational width of the present moment may seem a self-evident consequence of time's passing, the present flowing into the future even as it also simultaneously recedes into the past. That, however, is also the problem and the paradox, as Merleau-Ponty points out: "My present outruns itself in the direction of an immediate future and an immediate past and impinges upon them where they actually are, namely in the past and in the future themselves."[3] How can the present "actually" touch the past and the future? Or as the neurophenomenologists Shaun Gallagher and Dan Zahavi ask, "How can we be conscious of that which is no longer or not yet?"[4] As Husserl notes, "It belongs to the essence of lived experiences that [the present moment] must be extended in this fashion, that a punctual phase can never be for itself" (70). The contradiction "punctual phase" captures the paradox of lived time, a contradiction that prompted Husserl to call temporality oxymoronic, like "wooden iron."[5] The coherence of lived time as an integrated structure of differences ("punctual phases") is a fundamental, self-evident aspect of experience, but it is also a paradox that begs phenomenological and neuroscientific explanation.

This paradox is crucial to understanding reading and interpretation, which are phenomena that happen in time. As Iser observes, "It is impossible to absorb even a short text in a single moment."[6] Durational time characterizes both reading and life. That is one reason why, as Iser notes, "the reader experiences the text as a living event": "reading has the same structure as experience," because "meaning itself" in both domains "has a temporal character."[7] In reading as in life, consistency building and the to-and-fro movements of the hermeneutic circle, better called a "spiral," are temporal processes that manifest the paradoxes of lived time.

What are the neural correlates of time passing? What neurological processes underlie the paradoxical interpenetration of past, present, and future in the experience of duration? Neuroscientific explanations of how neurons fire and how assemblies of cells in different regions of the brain reciprocally interact are consistent with phenomenological descriptions of the lived experience of the "specious present," and these explanations provide a material, biological basis for phenomena that might otherwise seem mystifying and mysterious. The temporality of cortical processing helps to explain how the brain negotiates the competing claims of constancy and openness to novelty and multiplicity. It also suggests further reasons why the contrasting experiences of harmony and dissonance have been perceived to have aesthetic (not to mention pragmatic) value. The playfulness of the brain, as evidenced in its responsiveness to these and other aesthetic phenomena, is inherently temporal. Explaining the brain's temporality is necessary to understanding how it can play.

The vital importance of the capacity to integrate disparate perceptual moments, a capacity we ordinarily take for granted, is perhaps best illustrated when it breaks down, as in instances of motion blindness (also called "motion agnosia" or "akinetopsia"). In a case frequently cited in the neuroscience literature, a woman identified as LM suffered a stroke in the motion-processing region of the visual brain and lost her ability to integrate what she saw. "For this patient," as Bernard Baars and Nicole Gage explain, "the world appeared to be a series of still snapshots, like living in a strobe-lit world."[8] She could not pour a cup of coffee, because she literally could not see it filling up and was suddenly surprised to find liquid spilled all over the table. She could not cross the street by herself, because cars she thought were far away were all at once right on top of her. Like other victims of motion blindness, she also had difficulty conducting conversations, because she could not read the lips of her interlocutors and had to do without this important visual clue to deciphering aural speech patterns.

The bewilderment caused by the disruption of LM's visual experience dramatically illustrates how bizarre our perception of the world would be if the present were a series of discontinuous "knife-edge" moments rather than "a duration, with a bow and a stern." Gallagher and Zahavi ask: "What if our ongoing, present experience lacked a temporal coherence? What if, for example, I was unable to keep the just previous moment of experience in mind long enough to write it down, or was unable to anticipate events in the next second? Would my experience make any sense at all?" (70). The answer, simply, is no. If LM's experience "made sense" to her, that is because the temporality of other sensory modalities was still intact and gave her ways of compensating for her blindness to motion. That is also no doubt why her experiences of visual discontinuity seemed surprising and peculiar to her. Simultaneity is not simultaneous with itself but merges into the past and the future, and that is a good thing.

According to Husserl, any moment is characterized by *retentional* and *protentional horizons.*[9] The horizon metaphor suggests the paradoxical boundedness of the present and its connectedness to the past and future. Like a horizon, the present offers a perspective that is limited in its view but points beyond its boundaries—to what we expect (across the protential horizon) based on what has been (the retentional horizon). The temporality of lived experience, for Husserl, is a manifestation of how the world presents itself to us incompletely, in *aspects* or *profiles* that fit together (or don't) as our experience unfolds. Phenomenologically, "a thing is always intuited as something exceeding that aspect of itself which is actually perceived; the thing, so to speak, is always beyond the perception of the thing."[10] The perspectives through and in which we perceive the world are bounded in ways that we assume will be completed through other views as experience continues across the horizons of the present moment.

The present has a retentional horizon because the past is always slipping away, even as we preserve an ever-changing sense of what it was. "A now-phase is thinkable only as the boundary of a continuity of retentions," Husserl explains. "I can re-live the present but it can never be given again" (55, 66). The past is given to us across this horizon in a series of profiles that vary as the perspective of the present shifts. The retentional horizon of the present is the lived, intuited experience of the immediate past. It is not the same as memory, but it makes memory possible. For Husserl, there can be memory of any kind only because the present moment includes an immediate, retentional sense of what has just been.

Unlike explicit acts of remembrance, the retentional horizon does not represent specific experiences but instead provides an intuitive apprehension of what has just passed. It is consequently narrower than the "specious present" as William James defined it, whose "nucleus is probably the dozen or so seconds or less that have just elapsed" (*Principles* 1:613). James refers to something like what has come to be called *working memory*, the recollections we can draw on most vividly and reliably for whatever processing task we are engaged in, usually only four to seven items (see Baars and Gage 8). Although not the same as the retentional horizon, working memory gives evidence of the changing and receding character of what we hold in our most recent store of recollections—the limits and variability of what is available for us to "work" with—and this passing away is a manifestation of the horizonal presence of the past. As Husserl explains, the past is always "running off" (*ab-laufen*). Consequently, he continues, "memory is in a continuous flux because conscious life is in constant flux and is not merely fitted member by member into the chain. Rather, everything new reacts on the old" (77). The retentional horizon of the present provides an ever-changing perspective on the past through and across which we remember it. Memory is fluid and changeable because of the interactions of past and present across the retentional horizon.

Analogously, as we project expectations about how the aspects given at any moment will complete themselves, the protential horizon bounding the present offers shifting views of the "not yet." As Evan Thompson explains, we know that "our consciousness always involves an open and forward-looking horizon," because "it is always possible in principle for us to be surprised" (319). Francisco Varela points out, however, that "protention is generically not symmetrical to retention," because we have not yet experienced the future as we have the past; the protentional horizon is, in his words, "an openness that is . . . indeterminate but about to manifest."[11] The paradox of protentions is that they are both distinct and open-ended. The experience of surprise shows that we had particular expectations that were sufficiently determinate that they could fail to be fulfilled, even if an anticipation of the future is only a projection that as such is never completely specified.

Retention and protention are similar, however, in that they do not pre-

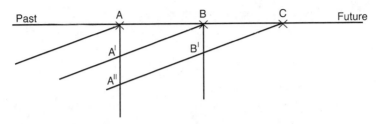

Figure 4.1. The profiles (*Abschattungen*) of passing moments. The horizontal line represents a series of present moments (*A*, *B*, *C*). The oblique lines represent profiles of these moments across the retentional horizon of successive later moments. The vertical lines represent the same moments in their different profiles over time. Adapted from Edmund Husserl, "Vorlesungen zur Phänomenologie des inneren Zeitbewusstseins," ed. Martin Heidegger, *Jahrbuch für Philosophie und Phänomenologische Forschung* 9 (1928): 440.

sent the past or the future fully and directly, in and of themselves, but only through a series of perspectives that are always changing. Consequently, Merleau-Ponty argues, "time is not a line" of distinct points but, rather, a "network of intentionalities" (417), an ever-moving array of horizons as the past and the future manifest themselves in a series of shifting profiles (see fig. 4.1). Just as we only know the world incompletely, through the ever-changing aspects in which it offers itself, so too are the past and the future (what has been and what is not yet) given to us horizonally and perspectivally.

The horizonal character of the moment has important implications for meaning creation. For example, as Husserl shows in his analyses of internal time-consciousness, we can compose sounds into a melody only because the present moment is linked horizonally to a network of past and future moments: "When the new note sounds, the one just preceding it does not disappear without a trace; otherwise, we should be incapable of observing the relations between the notes which follow one another" (30). Rhythm similarly only exists in and across time. The relation between beats in a syncopated rhythm, for example, is a horizonal temporal construct that we can perceive only because the present moment paradoxically includes the immediate past and future. As Tim Van Gelder rightly points out, "Auditory pattern recognition" is "a classic case of time consciousness" ("Wooden Iron?" 251).

It is also an important link between music and language. In his authoritative study of this topic, the neuroscientist (and musician) Aniruddh D. Patel notes that "there is evidence for overlap in brain processing of phrase bound-

aries in both domains," as well as in the recognition of syntactical and semantic patterns, including "melodic contours."[12] This should not be entirely
surprising, because as Lawrence M. Zbikowski points out, "Understanding
music is not simply a matter of processing auditory signals—it involves a
number of higher-order processes that humans use in a variety of ways to
structure their understanding of the world."[13] Reviewing the evidence on the
connections between music and language, Daniel J. Levitin concludes that
they "share some common neural resources, and yet have independent pathways as well."[14]

For example, he points out, musical syntax (the formal structure of music,
including keys and scales) is processed in Broca's area, and musical semantics (the meaningful associations that melody and harmony can convey) in
Wernicke's area, regions of the brain long associated with those dimensions
of language. Patients with lesions to Broca's area can understand meaning but
cannot form coherent sentences (a syntactical disturbance), whereas patients
with damage to Wernicke's area formulate fluent, grammatical, but meaningless sentences (a semantic deficiency).[15] Levitin's own experimental research
with fMRI technology has similarly suggested that temporal structure and coherence in music are processed in the same brain areas as language. Although
stroke victims can lose either music or language and retain some aspects of
the other, this indication of functional independence is balanced by powerful evidence that similar brain processes are invoked by both, because they
involve temporal acts of pattern formation.

A melody is not an objective entity but a developing temporal construct.
Sounds would be meaningless if they stood by themselves and could not be
joined perspectivally into an emerging sense of the pattern they compose.
This figure is never completely or simply present but is a retentional and protentional structure of relationships across our temporal experience. Rhythm
is similarly a structure in which the absences—the spaces between the
beats—are as important as the notes it joins because it is a figure-ground relationship. As Van Gelder points out, no "buffering" is required to understand
a melody: "In order to recognize a tune as such, we don't (as computational
models suggest) have to wait until the end of the tune. I hear the tune even as
the tune is playing" because "the system begins responding to the pattern as
the pattern that it is from the moment it begins" (258).

This point should be obvious to anyone who remembers the momen-
tous opening to Beethoven's Fifth Symphony: short-short-short-long, short-

short-short-long, . . . A more mundane example is the common experience of turning the radio dial and recognizing within seconds the musical genre identified with a particular station. Musical pattern is a temporally evolving part-whole relationship, and we project a sense of the gestalt we expect as soon as we can. This is one reason why the first sentences of great novels are often so memorable and so revealing (name that novel: "It is a truth universally acknowledged, that a single man in possession of a good fortune, must be in want of a wife"; or "Happy families are all alike; every unhappy family is unhappy in its own way" or "Someone must have slandered Josef K., for one morning, without having done anything truly wrong, he was arrested").[16]

Understanding is a temporal process of consistency building that utilizes the retentional-protentional horizonality of the present moment to project an anticipatory sense of pattern, and this expectation will in turn be modified, refined, or overturned as our experience unfolds. As Thompson explains, "Retention motivates protention, which affects retention, which motivates protention, and so on and so forth, in a self-organizing way that gives temporal coherence to experience" (361). The hermeneutic reciprocity of part and whole enacts itself in our experience as a to-and-fro play of anticipation and retrospection, with retention and protention refining and modifying each other. This play would not be possible if the lived moment were a "knife-edge" rather than a horizonal duration.

Reading is a temporal phenomenon that is similarly characterized by the creation and dissolution of patterns as we navigate our way through a text. Because "the whole text can never be perceived at any one time," as Iser explains, the reader occupies an ever-changing position—what he calls a "wandering viewpoint"—"at the point of intersection between retention and protention."[17] As we build provisional patterns of consistency and then revise them, the experience of reading entails "a process of continual modification [of meaning that is] closely akin to the way in which we gather experience in life" (*Implied Reader* 281). Reading involves "a continual interplay between modified expectations and transformed memories" (*Act of Reading* 111), an ongoing, reciprocal interaction between protentions and retentions across the horizons of the present moment.

As we read, Iser explains, we "view the text through a perspective that is continually on the move, linking up . . . different phases" of our experience and playing past perspectives off against each other in a "process of reciprocal spotlighting" (*Implied Reader* 280; *Act of Reading* 114). Because we never have the "text itself," but only shifting perspectives on it, the same text can be "concretized" (to use Ingarden's term) in different ways. Ingarden describes three different modes of "concretization," distinctive perspectives in and through which a work can present itself. We might simply experience the work with minimal reflection (the "aesthetic attitude"), or focus on the artistic object we believe we are experiencing (in a posture of "pre-aesthetic investigation"), or concentrate instead on the cognitive and emotional experiences we are having ("reflective cognition of the aesthetic concretization"), or, probably most typically, we may alternate among these different attitudes as we read, reviewing and assessing and revising our temporally unfolding experiences.[18]

These different attitudes are only possible because a text presents itself to us through variable temporal perspectives. One consequence of the temporality of reading, as Iser notes, is that "each concretization of meaning results in a highly individual experience of that meaning, which can never be totally repeated in its identical form. A second reading of the text will never have the same effect as the first" (*Act of Reading* 149). "The second time he [or she] is looking at the text from a different perspective. . . . Thus even on repeated viewings a text allows and, indeed, induces innovative reading" (*Implied Reader* 281). We can't read a text the same way twice. That is why we are sometimes disappointed when we return to a book or essay and find it not to be what we remembered (why did I include *that* on my reading list? I sometimes have asked myself the night before a class) and also why we sometimes are surprised to discover new meanings in a text we thought we knew well (there are some books I teach every year and am delighted to read and reread and reread yet again, and they seem different every time). In reading as in life, we cannot experience the same present moment twice, although we can remember it, but the present we remember will always be different at each recollection.

There is abundant evidence that brain processes are also durational. It would be very peculiar indeed if our lived experience of time were horizonal and recursive, while the neuronal mechanisms underlying it were pointlike and discrete. Francisco Varela, who pioneered the neurophenomenological study of time, points out that brain processes are characterized by "a frame or window of simultaneity that corresponds to the duration of [the] lived present," what he calls a "fusion interval," which is "the minimum distance needed for two stimuli to be perceived as nonsimultaneous."[19] For example, two lights flashing at an interval of less than 50 msec (recall that 1,000 msec [or ms] = 1 second) will be perceived as simultaneous, but they will appear to be sequential if the interval is more than 100 msec. When the interval is between 50 msec and 100 msec, the flashing lights will seem to move in the direction of the later light (an effect sometimes exploited by advertising in illuminated signs).[20] It is commonly understood that we locate sounds in the horizontal plane (to the left or right of us) by computing the *interaural time delay* between sound waves striking our ears, an interval that may be as little as 0.6 msec. This is a tiny instance of nonsimultaneous simultaneity without which we could not triangulate auditory direction. It is amazing (and a puzzle that neuroscience has not yet fully figured out) that the brain can compute such small temporal differences, because it takes neurons longer (1–2 msec) to fire. Even more astonishing, bats can distinguish time delays as small as 0.00001 msec.[21] The effect of apparent simultaneity—our unawareness of these temporal differences—is a phenomenon that Varela, Thompson, and Rosch call "perceptual framing": "Everything that falls within a frame will be treated by the subject as if it were within one time span, one 'now'" (73, 75). This "frame" is the lived, horizonal present.

The brain's "now" frame is relatively fixed biologically, but it can also vary within a particular sensory modality, as well as between different modalities, and it can be broadened by experience and training. For example, after noting that "two discrete sensory events in any modality can be integrated into a single conscious event only if they occur within 100ms of each other (approximately)"—their examples are "fast clicks, brief tones, visual flashes, [and] sensory taps"—Baars and Gage point out that "the 100-ms integration time does grow much larger in speech or music perception, or dance performance, where momentary events are interpreted in a much longer contextual framework" (289). Levitin cites experimental evidence that cortical areas associated with musical structure respond to stimulation within 150–400 msec and then are followed 100–150 msec later by areas associated with musical meaning, although the listener perceives no temporal lag (124). Even within the 100 msec frame there can be significantly different temporal response times. Recall from chapter 3 that Zeki reports experimental findings that "color is perceived before motion by about 80–100 ms"—"a trivial difference," he says, but "enormous" when compared with "the time taken for the nervous impulse to cross from one nerve cell to the next across the synapse, which lies between 0.5 and 1ms."[22]

Even beyond the 100 msec threshold, neuronal processes that can be measured discretely may be experienced as simultaneous because the brain integrates them into a meaningful pattern. Although, as Zeki notes, colors are perceived before locations and orientations, and "expressions on faces are perceived before their identity," he cites evidence that "over longer periods of time, in excess of 500ms, we do see different attributes in perfect temporal and spatial registration (the attributes are 'bound' together)."[23] The perceived "now" may vary in width, then, from 50 msec up to 500 msec and even longer, depending on the processing time needed by the brain to integrate different signals.

This is why Varela distinguishes between three different time scales of neuronal integration: the "1/10 scale" of basic sensorimotor and neural events (10–100 msec), the "1 scale" of "large-scale integration" (from fractions of a second up to 2–3 seconds, "the time it takes for a cognitive act to be completed"), and the "10 scale" of "descriptive-narrative assessments," a longer span in which explicit acts of representational memory may occur.[24] The 1/10 scale and the 1 scale correspond to the lived, horizontal "now," the width of the present moment, which can change according to the kind of cognitive experience that happens to transpire. Varela attributes variances in this span to two basic causes: "the intrinsic cellular rhythms of neuronal discharges," which may differ in particular sensory systems (different neuronal response rates in hearing or vision, for example); or "the temporal summation capacities of synaptic integration" between "brain regions," which are "interconnected in a reciprocal fashion" (an example would be the visual and aural pathways that combine in reading) and which take different amounts of time to complete their interactions (273, 274). As Zeki notes, "Binding between attributes takes longer than binding within attributes"—for example, more time is needed to integrate inputs from vision and hearing than to synthesize visual signals—but even within one system, like vision, the binding of similar features occurs faster: "the binding of color to motion occurs after the binding of color to color or motion to motion" ("Disunity of Consciousness" 217, 216). More complicated acts of cognition that bring widely dispersed cortical regions into relation (as in reading or listening to music or watching a dance performance) would consequently be expected to entail broader "now" frames than instantaneous events like flashing lights, which invoke only one sensory modality.

These discrepancies in neuronal processing time suggest that the brain is decentered both spatially and temporally. As Varela explains, "The relevant brain processes for ongoing cognitive activity are distributed not only in space but also over an expanse of time that cannot be compressed beyond a certain fraction of a second, the duration of integration of elementary events" (274). This divergence between the timing of a stimulus and its neuronal integration is a manifestation of the temporal split that characterizes not only human existence but all forms of life. The nonsimultaneity of the brain's cognitive processing is one aspect of life's inherent temporal imbalance. As Thompson explains, "Life is asymmetrically oriented toward the future" because "life's basic 'concern' is to keep on going" (362). Futurity is not, as existentialists in the Sartrean or Heideggerian tradition might claim, the exclusive purview of human beings. Rather, the temporal instability that allows human existence to be concerned about what Heidegger calls its potentiality for being (*Seinkönnen*) is a manifestation of the gap that can never be completely closed between a stimulus and the neuronal and other biological activity that organizes the organism's response to it.[25] The extremes to be avoided are stasis (the complete simultaneity of the organism with itself that is death) and instability that never leads to integration (like the growth gone off the rails that is cancer). In cognitive terms, the challenge to the brain as the organ coordinating the human organism's response to its environment is to balance homeostasis (preserving stability) and adaptability to changing conditions. The temporal gap that characterizes its processing time is not the disadvantage it might seem to be, because it allows the brain to play in the ever-changing horizonal space between past equilibria and the indeterminacies of the future.

With more complex forms of life, like human beings, this gap is related to a variety of interesting and important phenomena. As the neuroscientist Antonio Damasio puts it, "We are probably late for consciousness by about five hundred milliseconds." This means that we are always trying to catch up with ourselves: "By the time you get 'delivery' of consciousness for a given object, things have been ticking away in the machinery of your brain for what would seem like an eternity to a molecule—if molecules could think. . . . And because we all suffer from the same tardiness no one notices it."[26] But this tardiness is why there can be temporal forms of expression that play with the discrepancy between our expectations and their fulfillment (or frustration), like music or literature. If we instantaneously processed phenomena and there were no horizonal width to experience, there could be no play between anticipation and retrospection, no provocation to form patterns in order then to have them modified, refined, or overturned. The temporal instability of cognitive processing is the neural correlate of various forms of doubling that are crucial to aesthetic experience as well as to other cultural phenomena.

These include the doubling of consciousness back on itself that constitutes self-consciousness, the enabling condition of philosophical reflection. As the Danish philosopher Søren Kierkegaard memorably declared, "We live forward, but we understand backward."[27] When we reflect, as Merleau-Ponty points out, "consciousness always finds itself already at work in the world": "our reflections are carried out in the temporal flux onto which we are trying to seize" (432, xiv). The temporal width of the lived present, the difference between protentional and retentional horizons, makes it possible for us to reflect on our experience by putting one moment in relation to another across the "network of intentionalities" that is lived time, but the gap between the present of reflection and the past it can only re-present but not experience again also prevents us from ever attaining complete and transparent self-knowledge. This philosophical paradox is not entirely reducible to the temporality of brain processes, any more than music or literature are. But the gaps and instabilities in the brain's mechanisms of integration—the neural correlates of the phenomenological horizonality of time—are the neurobiological underpinnings of the distinctive temporalities of art and philosophy. Only because the brain is not simultaneous with itself can there be the modes of meaning creation that constitute music, literature, philosophy, and, indeed, culture.

Narrative, for example, is a literary form that is based on the brain's nonsimultaneity. The gap between *story* and *discourse*, the events of a tale and how they are told, which creates the possibility of narrative, is a manifestation of the temporal disjunctions at the neuronal level, which make life a perpetual process of catch-up.[28] Traditionally, one major purpose of narrative is to organize our experience of time, to arrange life into patterns with a beginning, a middle, and an end that create, in Frank Kermode's classic terms, "concord" out of the "discord" of temporal flux.[29] This narrative work of temporal synthesis would not be possible—or necessary—if experience were temporally at one with itself. The difference between the time of the tale's events and the time of their telling also allows for all sorts of innovative play with narrative time, which Gérard Genette calls *anachronies*—the flashbacks (his term is *analepses*) or flash-forwards (*prolepses*) that may interrupt the continuous movement of narration from the beginning of the story to the end.[30]

These interruptions in temporal continuity can have various effects on the reading experience. Joseph Conrad's sometime collaborator the novelist Ford Madox Ford, provocatively declared that in their view, "what was the matter with the Novel, and the British novel in particular, was that it went straight forward, whereas in your gradual making acquaintanceship with your fellows you never do go straight forward."[31] Although such temporal coherence is, strictly speaking, unrealistic, the so-called good continuation Ford criticizes may actually facilitate verisimilitude by assisting the reader's ability to build consistent patterns, thereby encouraging immersion in a lifelike world that seems to have the stability we take for granted in our everyday dealings with people, places, and things. This continuity disguises the temporal processes of understanding it manipulates, the interplay between anticipatory projection and retrospective modification through which the protentional and retentional horizons of experience mutually modify each other. Narrative disjunctions that interrupt consistency building may interfere with immersion in an illusion but can in turn promote reflection about the how of storytelling and the how of our temporal engagement with the world.

The temporal experiments of modern novels like Faulkner's *Sound and the Fury*, for example, or Conrad's *Lord Jim* or Ford's *The Good Soldier* refuse the reader's expectation of coherence in order to call attention to the disjunctions that make life a matter of living forward and understanding backward. With temporally disjunctive texts, realistic narrative immersion is sacrificed for epistemological reflection.[32] These experiments with narrative time would not be possible, however, without the disjunctions in experience that are the correlate of the nonsimultaneity of the brain's neuronal processes. Their aesthetic effect is to lay bare some of the epistemological consequences of these gaps and discrepancies that we do not typically notice because the syntheses of cognitive integration cover them up.

The temporality of brain processes begins with the way individual neurons fire.[33] An *action potential* is generated when electrically charged ions enter or exit a cell through channels that open and close, causing changes in the charge difference between the inside and the outside of the neuron. The resulting wave of *depolarization* and *repolarization* of the cell travels down the axon, the extension of the neuron that connects at the *synapse* with another neuron. The movement of an action potential is similar to the way a flame progresses along the fuse of a firecracker.[34] Like a wave, an action potential has a certain amplitude and frequency, and neurons can have unique electri-

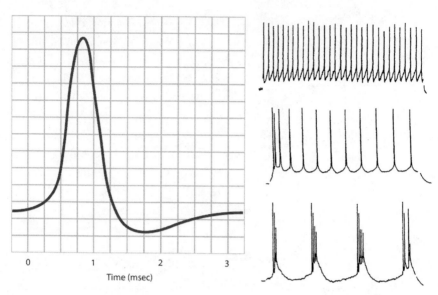

Figure 4.2. A, The characteristic "wave" of a single action potential. Drawing by Maggie Buck Armstrong. B, The "signatures" of three different neurons, with different amplitudes and frequencies. Adapted by permission from Ariel Agmon and Barry W. Connors, "Correlation between Intrinsic Firing Patterns and Thalamocortical Synaptic Responses of Neurons in Mouse Barrel Cortex," *Journal of Neuroscience* 12.1 (1992): 319–29.

cal signatures (see fig. 4.2). It takes a certain amount of time to generate an action potential (about 1–2 msec), and then there must be a certain resting interval, or *refractory period*, before the cell can fire again. How quickly the action potential can move along the axon to the synapse, where it is transferred to another neuron, also varies according to the thickness of the axon and how much *myelin* insulates the passageway (thicker, better insulated axons provide faster conduction). As Mark Bear, Barry Connors, and Michael Paradiso explain, "The frequency and pattern of action potentials constitute the code used by neurons to transfer information from one location to another" (76). This "code" is an electrochemical wave that can carry information not only because a charge is either "on" or "off" (either a neuron fires or it doesn't) but also because of its temporal characteristics (how often, how quickly, and how strongly it fires, a pattern that defines its signature).

Things become more complicated as neurons interact. As Baars and Gage explain, "Single neurons are fast-charging and fast-discharging electrical wave generators. Circuits of neurons oscillate in more complex patterns"

(247). This area of neuroscience is somewhat controversial and not as well established as one might wish. One important obstacle to understanding brain rhythms is the imprecision of electroencephalogram (EEG) technology, which measures electrical activity in the brain through electrodes placed on the scalp (or, in some animal experiments, directly beneath the skull on the surface of the cortex). As Daniel Levitin explains, "The EEG is exquisitely sensitive to the timing of neural firings, and can detect activity with a resolution of one thousandth of a second (one millisecond). But it has some limitations. . . . Because the electrical signature generated by a single neuron firing is relatively weak, the EEG only picks up the synchronous firing of large groups of neurons, rather than individual neurons. EEG also has limited spatial resolution—that is, a limited ability to tell us the location of the neural firings" (126). Hence the quip attributed to the neuroscientist John Nicholls that using an EEG to understand the brain is like trying to understand traffic patterns in Los Angeles by analyzing the smog—indirect and not unrelated but not highly accurate.[35] Imaging technologies like fMRI, which locate brain activity more precisely, have comparatively slow temporal resolution; it may take as long as several seconds for blood flow to increase sufficiently in the active region to light up the detector.

There is considerable evidence that the waves generated by neurons form into rhythms that may be crucial to how different regions of the brain coordinate their activities.[36] Bear, Connors, and Paradiso caution that "exactly how the parallel streams of sensory data are melded into perception, images, and ideas remains the Holy Grail of neuroscience" (421). This is known in neuroscience as "the binding problem." One proposal, they report, is the hypothesis that "neural rhythms are used to coordinate activity between regions of the nervous system": "by momentarily synchronizing the fast oscillations generated by different regions of cortex, perhaps the brain binds together various neural components into a single perceptual construction" (592). According to Baars and Gage, "Synchrony is a pervasive feature of the brain, apparently to coordinate patches of neurons at different locations. Indeed, synchrony appears even when neurons are cultured in a laboratory dish, or when a thin slice of the thalamocortical core is kept viable by growth factors. Synchronous activity appears to be a self-organizing feature of neurons in their natural environment" (252). By synchronizing their periodicities, brain rhythms may make possible "temporal coding" between regions, whether between nearby layers of the cortex (as in the rear visual area) or between far-flung, disparate areas like those connected during reading or listening to music.

For example, the brain's response to smell is apparently organized both spatially and temporally.[37] Even though the stimuli themselves are simply different chemicals and are not intrinsically spatial, smells are encoded on *olfactory maps*, topographical patterns in the neural anatomy of the olfactory bulb that register neuronal responses to odors. When we smell something, spatial and temporal information is combined to identify the particular odor. It matters not only which population of neurons fire on the olfactory map but also how their collective action potentials "spike"—with what temporal sequence, rhythm, and pattern. In one experiment, neuroscientists figured out how to disrupt the synchrony of honey bees' response to scents; the result was that the bewildered insects could no longer discriminate between similar odors, although they could still tell the difference between broad olfactory categories.[38] The experience of smell triggers neuronal responses at different places on the olfactory map, and the brain further refines the meaning of these reactions according to the temporal synchronies between their discharges. Temporal patterns of response coordinate and thereby interpret the spatial patterns of neuronal maps.

This is apparently a general feature of brain functioning. As Damasio explains, "Besides building rich maps at a variety of separate locations, the brain must relate the maps to one another in coherent ensembles. Timing may well be the key to relating."[39] Cerebral rhythms in temporal synchrony may be part of the solution to what Thompson calls "the large-scale integration problem."[40] Varela explains that "for every cognitive act, there is a singular specific cell assembly that underlies its emergence and operation" (274). In these assemblies, brain regions are linked reciprocally, with populations of neurons exchanging charges back and forth and generating oscillating brain waves, which further coordinate their interactions. When we listen to music at a concert or watch a music video, for example, regions of the brain interact from the far corners of the cortex—auditory neurons in the midbrain, motor and sensory areas across the central sulcus as we tap our feet or recall playing an instrument, the visual cortex as we coordinate what we see and what we hear, and areas of the cerebellum and the amygdala as we respond emotionally.[41] These cortical regions can only be integrated into coherent patterns of response if their neuronal discharges can be aligned, and the phase synchrony of their oscillations would provide such a mechanism of coordination (although the experiments necessary to chart this in detail remain to be done).

These assemblies come and go, in a cycle of "excitation" and "relaxation" that exhibits a particular periodicity. This rhythm is a natural property not only of single neurons but also of collections of brain cells. It is the neural correlate of the lived experience of time passing. As Varela explains, the synchronization coordinating a population of neurons is "dynamically unstable and will constantly and successively give rise to new assemblies," and "the fact that an assembly of coupled oscillators attains a transient synchrony and that it takes a certain time to do so is the explicit correlate of the origin of nowness" (283).[42] After an assembly is synchronized through a wavelike pattern of oscillatory excitation, it relaxes and must form again—or be replaced by another assembly. This pattern of phases corresponds neurologically to the horizonality of the passing moment that Husserl describes. That cell assemblies form and then dissipate is a natural consequence of how brain waves peak and subside. This periodicity makes any integration provisional and subject to change, modification, and replacement by another assembly. It allows for the formation of coherent patterns but prevents them from becoming rigidly fixed, once and for all, and is thereby an important temporal mechanism for balancing the conflicting claims of constancy and openness to novel stimuli.

The brain is a complex ensemble of multiple cell assemblies, each oscillating according to a different rhythm, and not a single, monolithically unified structure that is fully synchronized and in lockstep. The brain's multiplicity is more like a cocktail party, with different and changing patterns of simultaneous conversation, than a chant in unison at a football stadium; and it is consequently richer in information.[43] One conversation may momentarily take center stage, but it may then recede and give way to another as attention shifts, or as one activity takes over from another, or as a particular sensory modality becomes dominant (what is that awful smell? what is that crashing sound in the kitchen?). The simultaneous coexistence of multiple cell assemblies at different wavelengths makes it possible for the brain to multitask. As Colin Martindale explains, it is better to think of the brain as a "neural network" of "massively parallel processes," all "doing whatever they do simultaneously," than as a "serial" or "executive processor," making one decision after another in a linear manner.[44] Attention can allow only one assembly at a time to come into focus, as when we alternate between seeing a rabbit and a duck in the gestalt-shift experiments, but the brain can regulate several functions simultaneously beneath conscious awareness (e.g., as we listen to music while we walk and think about the next paragraph we need to write), and the brain's decentered structure as a multitasking ensemble of parallel processes is both more efficient and more flexible than a linear organization would be.

The electrical rhythms recorded in the brain range from the low-frequency oscillations (measured in hertz [Hz], or cycles per second) associated with sleep, called Delta waves (less than 4 Hz), through Theta (3.5–7.5 Hz), Alpha (7.5–13 Hz), and Beta (12–25 Hz) waves, which signal various forms of waking attention, up to Gamma waves (26–70 Hz), which are thought to reflect the most active exchange of information over long cortical distances (see fig. 4.3). Although complex mathematical methods known as *Fourier analysis* can disentangle some of the dense, overlapping wave forms recorded by EEG measurements, our understanding of the linkage between particular brain activities and specific wavelengths is at best imprecise (in part because, remember, the EEG is an intrinsically blurry and indirect representation of brain activity). There may also be considerable variation in the frequencies of oscillation at which any given cell assembly can synchronize, just as the same music can be transmitted at different frequencies on the radio dial. To take this analogy further, it may indeed be the case that some frequencies in the brain act as "carriers" to others, just as a radio signal "carries" various frequencies of voice or music. According to Baars and Gage, "Brain waves often interact with each other, with slower rhythms tending to group faster ones," and they compare this organizational work to the way in which "a fundamental frequency (the number on your radio dial) acts as a carrier wave for faster changes in frequency or amplitude, which reflect the voice or music signal" (248, 262). If your brain is like a radio, that is because the physics of wave oscillation allows it to code its multiple, simultaneous functions in a way that provides both order and flexibility.

There are two conditions in which the brain does globally synchronize: sleep and epileptic seizures; and comas, general anesthesia, and other unconscious states are similar. The EEG of someone losing consciousness during an epileptic seizure shows global, hypersynchronized slow wave patterns similar to, but more jagged than, the synchronized slow waves of sleep (see fig. 4.4). It is thought that the hypersynchrony of sleep is protective, warding off consciousness so that the brain can rest and consolidate the results of the day's activities, whereas epileptic hypersynchrony interferes with normal func-

Delta

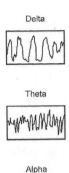

"Delta" waves (<4 Hz) are the slow, hypersynchronized waves of deep, unconscious sleep.

Theta

"Theta" waves (3.5-7.5 Hz) are associated with states of quiet focus, as in meditation, and with short-term memory retrieval.

Alpha

"Alpha" waves (7.5-13 Hz) are produced by the synchronous firing of large assemblies of neurons. They are also associated with awake relaxed states.

Beta

"Beta" waves (12-25 Hz) are associated with normal waking consciousness, focused attention, and anxious thought.

Gamma

"Gamma" waves (26-70 Hz) are produced by many conscious activities and are believed to connect cortical and subcortical regions of the brain. They also appear during dreaming in "rapid eye movement" (REM) sleep.

Figure 4.3. Brain-wave frequencies and functions. Adapted from Bernard J. Baars and Nicole M. Gage, *Cognition, Brain, and Consciousness*, 2nd ed. (Amsterdam: Elsevier, 2010), 107 (table 4.1). Drawings by Maggie Buck Armstrong.

tioning. This may seem paradoxical, as Baars and Gage point out. Whereas "synchronized brain rhythms allow widely separated regions of the brain to work together," they note, too much synchrony is apparently a bad thing: "the electrical storms of global hypersynchrony disrupt ordinary brain functions" and prevent even "normal survival activities" from being carried out (246).[45] The explanation of this paradox is that "normal cognition requires selective, local synchrony among brain regions"—"highly patterned and differentiated" oscillatory patterns, in which "synchrony, desynchrony, and aperiodic 'one-shot' waveforms constantly appear and disappear"—as opposed to the too-uniform global lockstep of hypersynchrony (246). In both sleep and epileptic seizures, hypersynchrony is like noise, then, but noise of different kinds, analogous to the distinction between the white noise that blocks out disturbances to peaceful rest and the loud, disruptive noise that interferes

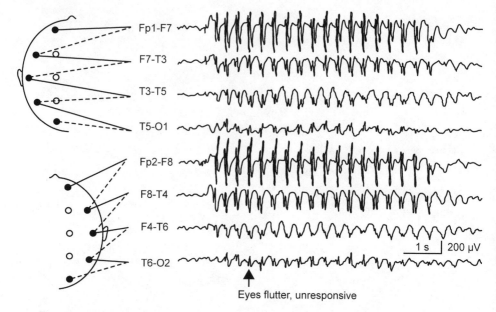

Eyes flutter, unresponsive

Figure 4.4. Hypersynchrony of epilepsy and sleep. EEG showing hypersynchronized brain activity during epileptic seizure of a seven-year-old girl. Sleep patterns are similarly hypersymmetrical but less jagged and can show long upward and downward waves. Reproduced by permission from Hal Blumenfeld, "Consciousness and Epilepsy: Why are Patients with Absence Seizures Absent?" *Progress in Brain Research* 150 (2005): 274 (fig. 2).

with attention, concentration, or the exchange of information. In both cases, however, the paradox is that these kinds of noise are not chaos or disorder but hyperorder, an excess of structure and coherence, which in other instances are characteristics of meaningful experience.

This paradox helps to explain why both harmony and dissonance can be pleasurable, meaningful aesthetic experiences and why both are different from noise. Harmony is not lockstep hypersynchronization but a more or less complicated structure of differences, a pattern that is coherent but not uniform (recall that Ingarden's classical aesthetic refers to a "*polyphonic* harmony of value qualities").[46] In music or literature that is harmonious, the patterns typically shift and develop and modify themselves over the course of the performance or a session of reading, in a manner parallel to how the neuronal cell assemblies underlying these experiences provisionally organize, dissipate, and re-form. Harmony can be stimulating and enjoyable rather than stultifying because, unlike sleep or a coma, it is differentiated, replete with multiplicity, and temporally variable. The harmonies of art resonate with, stimulate, reinforce, and restructure the synchronies of brain activity.

We can take pleasure not only in the recognition and inculcation of pattern, however, but also in its disruption, and this too is consistent with the temporal rhythms of brain activity. Aesthetically significant dissonance is not simply the hypersynchronic negation of differentiated, patterned meaning creation that destructive noise represents—analogous to the shutting down of flexible, multiple brain-wave activity when an epileptic seizure takes over. Dissonance that is meaningful (and perhaps pleasurable as well as useful) is equivalent to the phase scattering of synchronous organization that keeps the brain flexible and that opens the way to new forms of assembly—or, put differently, that shouts down one particular dominant conversation in the brain's cocktail party in order to make room for other voices to speak and other concerns or activities to take center stage. Dissonance too, then, can be part of the pattern of assembly and disassembly that gives brain activity its temporal, rhythmic, wavelike structure. In all of these ways, the to-and-fro play of harmony and dissonance is made possible by and is an experiential manifestation of the oscillating synchronization and desynchronization of neural assemblies.

Over time, experiences of reading can have effects on the brain, but what these are and what aesthetic and pragmatic value they may have are, interestingly, matters of some controversy. One key set of questions has to do with the consequences of repeated exposure and habit formation, which (as I have noted in previous chapters) such aesthetic theorists as Victor Shklovsky and Hans Robert Jauss denounce for dulling our response to art and life.[47] Following this tradition, it is common for contemporary cognitive literary theorists to identify *literariness* with *defamiliarization* or *dehabituation*. David Miall argues, for example, that "literature is dehabituating; that is, it invites us to consider frames for understanding and feeling about the world that are likely to be novel, or at least, unfamiliar."[48] The evolutionary literary critic Brian Boyd similarly claims that defamiliarization has pragmatic and aesthetic value because "habituation upon repetition occurs in any organism with a nervous system: *any* stimulus gradually ceases to arouse."[49] These claims deserve skeptical scrutiny, however, in light of the findings of neuroscience on habit formation. The experimental evidence for privileging dissonance over harmony is not at all clear, because habit and repetition can have widely divergent effects.

Despite the dangers of a dulled sensibility, habit-producing repetition also has its values. There is even a kind of pleasure that the formation of habits makes possible. According to the philosopher-neuroscientist Alva Noë,

> That novices and experts have qualitatively different manners of involvement with what they are doing has also been confirmed by neuroscience. It has been shown, for example, that highly trained experts—musicians, athletes, etc.— show a decrease in the overall level of brain activation when they are engaged in the performance of their skills compared to beginners. In a way, it is almost as if the better the player, the less there is for the brain to do! For the experienced player, the task takes over.[50]

It is not that the novice is having more fun than the expert because his or her brain is firing more intensely—quite the contrary. These results recall Gadamer's argument that in some instances the game can take over from the consciousness of the players when they become so immersed in the back and forth of their exchange that they experience a kind of momentary transcendence.[51] The immersion of effortless reading is not necessarily mind-dulling escapism, then, but may be a deployment of expertise that is inherently pleasurable and may even make possible experiences of self-forgetfulness that feel like self-transcendence. Habit-forming repetition is inherently ambiguous and can have opposite consequences, leading to either a reduction or a heightening of responsiveness. Not necessarily a bad thing, the unconsciousness of habituation may be a neuronal foundation of aesthetic joy.

Reading is a kind of "skillful coping" (to borrow Thompson's useful phrase [313]) that only develops through repeated practice. It takes many years of habit formation to become an expert reader who can take pleasure in Henry James, Virginia Woolf, or James Joyce. Once again Noë's commentary is insightful:

> I suspect that there is something good, something healthy, about shaking things up by disrupting one's habitual regimes. But I entirely reject the idea that it would be best to be entirely habit-free (as if that were even possible!). . . . You need habits of thought and behavior in order to be decisive and deliberate, for habit is the foundation of skill. . . . Habitual modes of thought and behavior are themselves frequently expressions of intelligence and understanding, even if they are spontaneous, automatic responses to things. (118–19)[52]

These points may not be controversial, but they are worth remembering. As a counter to the tendency in literary theory to inordinately value experiences of disruption, transgression, and disorientation, we should keep in mind that dehabituation has meaning and utility only against a background of estab-lished habits and conventions, which it may modify but cannot completely destroy without inducing paralysis. The brain needs both the constancy of habit and the flexibility of disruption. Learning involves not only the estab-lishment of smooth pathways and fast-responding connections in the brain through the reinforcement of repeated experiences but also the disassembly and reformation of patterns in response to novel stimuli. Aesthetically as well as neuroscientifically, it is a mistake to emphasize one pole to the exclusion of the other.

These speculations about the neuroaesthetics of habit are reinforced by experimental work with a humble member of the animal kingdom that has been a much-studied model of the neurobiology of learning. The classical neuroscientific experiment on habituation and sensitization in the lowly sea slug (*Aplysia californica*) is curiously and importantly ambiguous on the neu-ronal consequences of repeated stimulation in ways that have interesting im-plications for the contradictory responses of human beings to harmony and dissonance. In research that won a Nobel Prize, Eric Kandel and his team re-peatedly squirted water into the gill of the sea slug in order to figure out why this gradually deadened its withdrawal reflex. They then stimulated the slug's head with electricity and caused a sensitization of the response—that is, a quickened and exaggerated gill withdrawal (see fig. 4.5). The neurochemistry of the two opposite reactions is very complicated and need not concern us here (although that is what won Kandel his Nobel), except to note that in habituation there is a decrease and in sensitization an increase in the produc-tion of action potentials.[53] Noë usefully summarizes these results as follows: "Repeated harmless touching causes the strength of the connection between the sensory cells and the motor cells to weaken. . . . (Think of the way you don't feel your clothes.) Sensitization is the reverse process. Painful contact produces a strengthening of synaptic connection between sensory input and motor output. The snail learns; it remembers; it modifies its behavior in light of this learning" (92).

It is an interesting, nontrivial question, however, whether the poor sea slug perceived having water squirted on its gill as harmless and the electric stimulation painful. To invoke the term oft associated with Thomas Nagel's essay "What Is It Like to Be a Bat?," the "qualia" of the sea slug's experience is not something we can know.[54] The important point for my purposes, how-ever, is that both experiences disturbed the organism's relation with its envi-

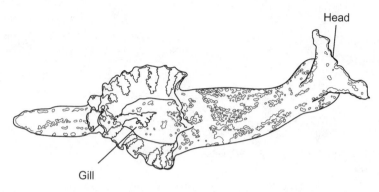

Figure 4.5. Gill-withdrawal reflex in the sea slug. Drawing by Maggie Buck Armstrong.

ronment, but with opposite results, dulling the reaction in one instance and sensitizing it in the other. A disturbing stimulus was repeated, but the repetition reduced responsiveness in one case and increased it in another.

In a further complication, sea slugs have also been shown to be amenable to "classical conditioning." That is, the gill-withdrawal reflex can be induced by contact to another part of the snail's anatomy through repeated associative stimulation (see Bear, Connors, and Paradiso 768–71). It is unclear whether this should count as habituation or sensitization. It is probably a different mechanism, the formation of a habit by deadening one response (the reaction ordinarily provoked by stimulating the area in question) and inducing another (the new, "conditioned" response). But this ambiguity is further evidence that the repetition of a stimulus is not by itself decisive as to how the organism will react. Habituation, sensitization, and habit formation are all different forms of habitualization that repeated experiences can bring about. Repetition can be either dulling, sensitizing, or habit producing depending on a variety of factors having to do with the kind of stimulus, the organism's characteristics, its history of previous stimulation, and the context of the interaction.

As is often the case, this experiment raises as many questions as it answers: Why do organisms respond so differently to repeated stimulation, and what are the implications for art? Is the repeated exposure to literature habituating or sensitizing, and can this be determined by a work's qualities—whether, for example, it is primarily harmonious or dissonant? Is a symphony by Haydn or Handel, say, more habituating and less sensitizing than a modern piece by Stravinsky or Schoenberg? Are the habits we form as readers intrinsically dulling, or can they condition us to modes of responsiveness that are more like sensitization? One difficulty here is that what counts as harmless or painful (either for an organism in a laboratory experiment or for the recipient of an aesthetic experience) is not an absolute but is contingent on a variety of factors, including the sensibilities of the organism (or of the particular reader or listener), and these in turn are a result of previous experiences that are themselves characterized by repetition (what music we perceive as harmonious or dissonant can vary widely depending on what we are accustomed to, as the parent of any teenager knows).

This is one reason why harmony can be just as sensitizing as dissonance can be—and why dissonance that tips over into noise can be dulling and habituating. The pain of a dissonant disturbance can prompt withdrawal unless it can be integrated, just as the harmony of a classical work can sensitize the recipient's response if it can promote a recognition of previously unnoticed differences. Repeated exposure to Haydn or to Stravinsky is just as likely to lead to boredom or to heightened appreciation—there is no telling in advance. The aesthetic moral of the sea-slug experiment is that there is no neuroscientific reason to favor one kind of aesthetic over the other; rather, both habituation and sensitization are neuronally based responses that can be triggered by repeated experience. Depending on the organism, the particular stimulus, and the context, harmony and dissonance can be either habituating or sensitizing. They are not intrinsically one or the other.

The workings of habit suggest that memory and learning do not occur at a single site but are distributed throughout the brain. As Gallagher and Zahavi explain, "Memory is not a single faculty of the mind" but entails "a variety of distinct and dissociable processes" (70). Some localization seems to characterize memory, so that *declarative memory*, what we can explicitly recall, is associated with the hippocampus, and *procedural memory*, what we remember how to do, is associated with the striatum, but even these links are not fixed and exclusive (see Bear, Connors, and Paradiso 725–59). There is considerable evidence that memory occurs where learning happens. For example, in one oft-cited experiment, brain scans showed different parts of the cortex lighting up when bird watchers and auto buffs saw the objects of their special expertise, suggesting that their memory sites were closely linked to their processing sites.[55] Their brains fired differently in response to birds and cars because their memories of these special fields, the storehouses of their expert knowledge, were part and parcel of their cognitive experiences. This is consistent with Hebb's proposal that the cortex is constantly rewiring itself in response to our experiences. Memory traces, or *engrams*, are created where processing occurs through the repeated formation of analogous cell assemblies and the strengthening of the reciprocal neuronal connections underlying them. An engram is a neurological habit.

The "skillful coping" of reading may be processed through the letterbox area of the visual cortex, but it also draws on learning and memory sites across the brain. Reading deploys memory not only through the recollection of specific contents but also in the very way we build consistent patterns and fill in indeterminacies based on past experiences with literature and life. The links between memory and processing in the brain facilitate these interactions. The way that repeated experiences of processing form patterns of neural assemblies helps to explain how particular reading practices can become inculcated, even to the point where readers with opposing interpretive conventions may find different meanings in the same text.

This is at least in part what happens neurobiologically during the interpretive conflicts that are characteristic of the humanities. Like the brains of ornithologists and auto buffs, the brains of psychoanalytic and Marxist critics would be expected to fire differently, not only because of what they know but also because of how they use their expertise. Our past experiences of the world, including the knowledge we have acquired and the beliefs we have come to adopt about literature and life, can make us read differently if they have become wired in our patterns of cerebral processing.[56] Members of different interpretive communities—deconstructionist or New Historicist, feminist or queer theorist, cultural critic or formalist—will have different engrams because of the memories they have formed from their histories of reading. These neurological structures are not only the traces of past processing but also the habits that come into play in present and future cognitive acts. Although there are limits to the brain's plasticity, and all of our brains have many structures and processes in common, the wiring of any two brains will differ because of the connections established by past cortical syntheses (neurons firing together, wiring together), and that is why two readers with different histories may respond very differently to the same novel or poem.

How much historical variation is there in how readers have read? The eminent book historian Robert Darnton complains that many reader-response theorists "seem to assume that texts have always worked on the sensibilities of readers in the same way. But a seventeenth-century London burgher inhabited a different mental universe from that of a twentieth-century American professor. Reading itself has changed over time."[57] In order to evaluate this claim, we need first to distinguish different time scales. Analogous to Varela's three time scales of microcognitive events, there are important differences at the macrolevel in the history of changes in the brain. (To be clear, it is the analogy here that matters; I am not proposing that macrolevel calculations can be made that precisely parallel microcognitive temporality.) The scale with the broadest time range (analogous to the 10 scale on the microlevel) would include the long evolutionary history of the human brain's development, which produced its particular features and abilities, different from (and also similar to) those of other mammals, whose origins we share. Our capacity for language—what Pinker calls the "language instinct"—developed on this scale. The medium-range historical scale (the analogy here is to Varela's 1 scale) would refer to the shorter but still considerable period required for these cortical structures and processes to be repurposed for enduring cultural tasks that are handed down through education from generation to generation. What Stanislas Dehaene calls the "neuronal recycling" through which reading emerged several thousand years ago is a change on this medium scale. On the smallest temporal scale (the macrolevel analogous to the 1/10 scale of microprocessing) are the individual variations that can occur in the brains of taxi drivers or pianists because of the Hebbian rewiring that takes place due to their histories of repeated practical activity. If individual readers with varying personal histories and hermeneutic allegiances read differently, or if reading practices have changed historically during the relatively short period (from an evolutionary standpoint) after literacy was introduced, these are occurrences on this smallest of the three scales of macrohistorical change.[58]

Phenomenology's description of reading as a to-and-fro process of building consistent patterns is based on fundamental cognitive, neurological processes on the first and broadest of these temporal scales; these processes are then "recycled" for reading on the second, medium-range scale of the last 5,500–6,000 years. The hermeneutic processes of configuring part-whole relations and their temporal enactment in a back-and-forth mutual modification of protentions and retentions are integral characteristics of the cognitive workings of the brain that have a long evolutionary history that antedates the development of our ability to read. These brain functions would be pretty much the same in a seventeenth-century London burgher and a contemporary American professor of literature (or even a neuroscientist). If Walter Ong is right that "more than any other single invention, writing has transformed human consciousness," that fundamental change occurred several thousands of years ago, when visual capacities for invariant object recognition came to be "recycled" for the purposes of reading.[59] This change repurposed but did not biologically alter the meaning-making processes of the brain. Subsequent changes in the technology of textual reproduction, from print to the Internet, and in the extension of literacy beyond the privileged few to a mass public readership may have had profound social, political, and cultural implications, but the basic cognitive, neurological, and phenomenological processes underlying these developments have not changed.

We may read differently from generation to generation (the analogy to the 1/10 scale), but we do so by deploying the same processes that have been in place ever since humans developed the capacity to recycle their brains for reading a half-dozen millennia ago, a repurposing of cortical cognitive processes that have an even longer evolutionary history (the broadest macroscale). As the neuroscientist of reading Dehaene memorably declares, "We take delight in reading Nabokov and Shakespeare using a primate brain originally designed for life in the African savanna."[60] The phenomenological and neuroscientific account of reading that I have been offering here not only applies to how readers have read for as long as reading has been possible; it also explains how, through the vagaries of habit formation and cortical rewiring, these enduring, invariant processes make possible the sort of developments and differences that Darnton calls attention to.

How much can experiences of reading change individual behavior? This question has enormous, much-debated implications for the moral and political effects of literature. If, in the oft-invoked Horatian formula, the purpose of literature is to "please and instruct," what can neuroscience tell us about the capacity of our encounters with texts to change our lives? Among theorists of reading, the answers have varied widely, from idealists' praise for the humanizing and emancipatory powers of literature to skeptics' suspicion of its coercive role as part of the cultural apparatus for inculcating behavioral norms. Jauss summarizes the idealists' case: "The experience of reading can liberate one from adaptations, prejudices, and predicaments of a lived praxis in that it compels one to a new perception of things" and may thereby contribute to "the emancipation of mankind from its natural, religious, and social bonds."

A powerful recent statement of the skeptical position is offered by Nicholas Dames, who questions the conventional wisdom about the liberalizing moral and social impact of nineteenth-century fiction: "The Victorian novel was a training ground for industrialized consciousness, not a refuge from it," and "condition[ed] the physiological apparatus of the reader for the temporal rhythms of modernity" because the verbose three-decker "trained a reader able to consume texts" at the fast pace required by modern life.[61]

From a neuroscientific perspective, it is certainly possible for literature to change a reader's consciousness, as repeated experiences alter cortical wiring in a way that reflects our personal and cultural history, including our encounters with literary texts and other art forms. For example, Patel cites considerable experimental evidence that "music has the power to change the very structure of our brains, enlarging certain areas due to motor or perceptual experience" (401). Dehaene similarly observes that "the letterbox area [through which reading is processed] is not merely determined by visual stimuli, but also by the cultural history of the reader's brain," and he notes that "even in the adult brain, learning can still drastically alter neuronal connections" (95, 211). But the brain is complexly balanced between inertia and openness to novelty and variation, and there are limits to its capacity to shed old habits and develop new patterns of response. As Iser points out, a version of the hermeneutic circle characterizes the learning that reading may bring about, because "the old conditions the form of the new," even as "the new selectively restructures the old" (*Act of Reading* 132).

Existing neural patterns are set in motion in response to novel phenomena in reading and other forms of experience, and in the process the cell assemblies through which these phenomena are assimilated may establish new connections in the to-and-fro reciprocal exchange between cortical areas. This play can result in changes in brain structure, but whatever alterations occur will be based on prior patterns, and long-lasting developments do not usually happen globally or instantaneously.[62] Rather, the neurological consequences of reading are typically a result of repeated exposure to patterns that form habitual responses—patterns that, alternatively, a dissonant reading experience may expose and undermine.

It usually takes a long time for neural structures to get established, and a single reading of one particular book is unlikely to transform them, although a lifetime of working with one particular author, genre, or period or within one or another interpretive paradigm (analogous to a career playing a specific musical instrument) may in all probability have neural consequences. It also matters here that reading competes with many other activities and experiences in daily life to form our brains, and so the consequences of our encounters with literature are likely to be buffered and diminished, if not necessarily overwhelmed, by other nonliterary influences. Moreover, "literature" itself is not a monolithic category, and a lifetime of reading a variety of texts will push and pull readers in many different directions (and concomitantly form and re-form cortical connections in their brains in various ways, some mutually reinforcing, others contesting established patterns). These are all reasons to be suspicious of claims that literature has unidirectional moral or political effects, whether for good or for ill (and an explanation too, perhaps, of why humanists are not more humane than other folks despite all of the time we spend reading great books).

These many variable, unpredictable factors all suggest that brain structure alone does not fully explain how we respond to novelty and is not finally determining. The set patterns in the brain are in tension with its ability to form new combinations of neurons in response to different situations, and this playful imbalance between stability and instability leaves ample room for variation, that is, for the contingencies and unpredictabilities of our interactions with the world that may be experienced as evidence of something like human freedom. The brain can be habitualized, as the deterministic skeptics fear, but its patterns are more or less open to alteration, as the idealists hope.

In reading as in life, these habits are not only cognitive but also emotional. Indeed, as Thompson points out, "Cognition and emotion are not separate systems" but instead "interact with each other in a reciprocal and circular fashion, up and down the neuraxis" (371). The integration of cognition and emotion is a consequence of the brain's embodiment. As Damasio explains, "Normal mental states are invariably imbued with some form of feeling" because of the brain's relation to the body.[63] Emotions are ultimately based on what he calls "primordial feelings, which occur spontaneously and continuously whenever one is awake" and "provide a direct experience of one's own living body, wordless, unadorned, and connected to nothing but sheer existence" (*Self Comes to Mind* 21). These "primordial feelings" are, he explains, "spontaneous reflections of the state of the living body" and "are based on the operation of the upper brain-stem nuclei, which are part and parcel of the life-regulation machinery. Primordial feelings are the primitives for all other feelings" (101). They are the foundation for "the so-called universal emotions (fear, anger, sadness, happiness, disgust, and surprise)," which, he argues, "are present even in cultures that lack distinctive names for the emotions"—"unlearned, automated, and predictably stable action programs" that have "their origin in natural selection and in the resulting genomic instructions" (123). Other affective states, like those Damasio calls "the social emotions"—"compassion, embarrassment, shame, guilt, contempt, jealousy, envy, pride, admiration"—may be of more "recent evolutionary vintage, and some may be exclusively human" (125–26).

Once again the brain is divided and decentered, not unified and centrally controlled, inasmuch as two systems coexist within it that are the result of different periods of evolutionary history—the brain stem, which regulates basic bodily functions and is the source of "primordial feelings," and the cerebral cortex, where higher-level activity (both emotional and cognitive) occurs. Situated adjacent to the upper brain stem, "the thalamus serves as a way station" between the body and the cortex, Damasio explains, relaying signals back and forth between them in "moment-to-moment recursive loops" that "interassociate information" between "spatially separate neural sites, thus bringing them together in coherent patterns" (*Self Comes to Mind* 247–48) (see fig. 4.6). Although these interactions are "seamlessly integrated," what Damasio calls the "interplay" between brain stem and cortex brings into relation "two somewhat separate 'brain spaces'" that "point to different ages in brain evolution, one in which dispositions sufficed to guide adequate behavior and another in which maps gave rise to images and to an upgrade of the quality of behavior" (248–49, 153).

As a process of "skillful coping," reading is an embodied activity that sets in motion relations among all of these various sites, from the brain stem up through and across the thalamus to the higher cortical regions and back down again. As we read, our brain and our body interact, and that is why reading is both a cognitive and an emotional experience. Citing experiments employing fMRI technology, G. Gabrielle Starr reports that "parts of the brain's architecture that coordinate motion are also recruited by metrical writing. . . . Poems may make us wish to keep time, to move and imagine motion."[64] This finding is consistent with well-known evidence that when pianists listen to music, their motor cortex is activated in places that also light up when they are playing, an experimental result consistent with reports from many of them that their fingers move in response to piano music.[65]

Further indications of relations between the language-processing areas of

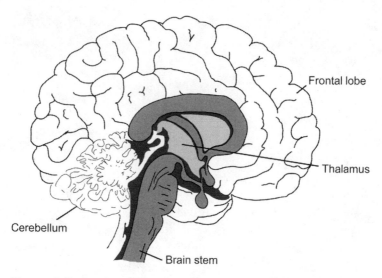

Figure 4.6. Brain stem, thalamus, and cortex. Centrally located on the upper brain stem, the thalamus mediates, or "interassociates," signals from the lower and upper areas of the brain. Drawing by Maggie Buck Armstrong.

the brain and the motor cortex have been provided by experiments suggesting that reading action words provokes activity in cortical areas related to the same kinds of physical movement.[66] These responses are so specific to our bodily habits that left- and right-handed subjects register responses to action verbs in opposite hemispheres of their brains.[67] In addition to offering evidence of brain-body interaction in experiences with art and language, these experiments suggest that the way literature and music organize time may inform and also reflect the brain's coordination of different temporal modalities (visual, auditory, and motor). The rhythms of poetic meter and music may help to structure the temporality of other brain-body processes.

Phenomenologically and neuroscientifically, emotion plays an especially important role in our orientation toward the future. The disposition of primary feelings is an organism's set toward the horizon of the "not yet." Clinical and experimental evidence of this connection between emotion and the protentional horizon is provided by Damasio's studies of patients with damage to the prefrontal sensorimotor cortex. He reports that their inability to form emotional dispositions based on their bodily experiences prevents them from thinking beyond the present moment and results in various irrational and sometimes sociopathic behaviors that reflect a lack of concern about or awareness of the implications of their actions. This "myopia for the future," as he calls it, is a result of their inability to inform their intact cognitive processes with emotionally based intuition.[68]

The explanation for this strong connection between emotion and futurity has to do with the peculiar, paradoxical quality of the protentional horizon, which, as Thompson points out, entails "a more or less definite" anticipation of the future that "is nonetheless open and indeterminate, for what it intends has yet to occur" (360). Recall that protentions lack the specificity of the retentional horizon, which provides a particular perspective on determinate events that have already happened. By contrast, Varela describes the protentional anticipation of the future as an "immanent affective disposition" or a mode of "readiness, an expectation about the way things in general will turn out" (303, 299). According to Thompson, the "forward trajectory of life and mind is fundamentally a matter of emotion" because "the protentional 'not yet' is always suffused with affect and conditioned by the emotional disposition (motivation, appraisal, affective tone, and action tendency) accompanying the flow" of experience (362). This is why Heidegger describes the "anticipatory structure" of interpretation (what he calls its "forestructure," or *Vorstruktur*) as a matter of our "mood" or "attunement" (*Stimmung*) and of our affective "state of mind" (*Befindlichkeit*). "In every case," writes Heidegger, "Dasein [human existence, lit. "there-being"] always has some mood. . . . A mood makes manifest 'how one is, and how one is faring.' In this 'how one is,' having a mood brings Being to its 'there.' "[69] Biologically as well as existentially, our mood, temperament, attunement, disposition, and readiness are all ways in which the paradoxical presence and absence of the "not yet" manifests itself on our protentional horizon.

There is abundant experimental evidence of the role of attunement, disposition, and readiness in gearing our expectations in ways that prestructure and consequently to a great extent predetermine what we perceive. Well-known effects of *subliminal priming*, in which visual stimuli appear and vanish so quickly that we do not "see" them but they still influence how we act, only occur because what we perceive is molded by our unreflective disposition. Gallagher and Zahavi report an elaborate and precisely controlled experiment in which the different states of "readiness" or "unreadiness" of subjects for particular stimuli resulted in markedly different response rates and patterns of brain activity. States of readiness produced reaction times in the 200 msec range, as opposed to 600–800 msec for unreadiness, and topographical EEG measurements over the surface of the scalp, even taking into account their inherent "smog-like" imprecision, suggested more rapid and more organized productions of synchrony in the response patterns of "ready" subjects.[70] Converse effects have also been demonstrated. For example, as David Miall reports, the foregrounding of particular words or sound patterns can slow down reaction times in ways that are measurable on the 1/10 scale (in a range of 162–354 msec per word) by interrupting the expected flow of sentence meaning.[71] Processing is faster when expectations are met and slower when they are not, and both effects demonstrate the influence of disposition and readiness on cognition.

These results are not surprising; they are consistent with generally acknowledged effects of priming expectations as well as frustrating them, effects that are typical in literature, music, and daily experience.[72] For example, Alvin Goldman reports an experiment in which subjects were first asked to write sentences from word lists associated with a particular trait or stereotype (like *gray*, *bingo*, and *Florida* for "elderly") and were then tested against a control group primed with neutral words. "Subjects primed with the *elderly* stereotype took longer to walk from the lab to the elevator"; subjects primed with "rude" words "interrupted the experimenter more quickly and frequently" than a group given a "polite" list; and subjects primed with terms about politicians were then prone to "longwindedness" and "wrote longer essays on nuclear testing" than a control group.[73]

Some effects of priming are less predictable and more peculiar, however. Noting that "expectation and anticipation may trigger strong imagery with perceptual effects," Gabrielle Starr reports curious experimental findings demonstrating that "imagined odors . . . may interfere with or alter our perception of actually present tastes": "Certain smells, like that of strawberry, make it easier for us to taste sugar. The imagined smell of strawberry does the same," but "visual imagery has no such effect."[74] Whatever the exact neurobiological explanation of these linkages turns out to be, they are dramatic evidence of the bodily basis of attunement. Different sensory modalities can reinforce each other because they embody dispositions toward the future. In all of these ways, understanding in life as in art is futurally oriented toward the horizon of the "not yet." The aesthetic effects of harmony and dissonance are achieved by manipulating these expectations, either to confirm or disrupt them (and usually both in some combination), and these effects are cognitive, emotional, and embodied.

The priming and frustration of expectations can set in motion what Damasio calls an "as-if body loop," which can have powerful emotional effects in both art and life. According to Damasio, "The brain can *simulate* . . . certain body states, *as if* they were occurring" (*Self Comes to Mind* 102, emphasis in original):

> Because our perception of any body state is rooted in the body maps of the somatosensing regions [of the brain], we perceive the body state as actually occurring even if it is not. . . . The as-if body loop hypothesis entails that the brain structures in charge of triggering a particular emotion be able to connect to the structures in which the body state corresponding to the emotion would be mapped. For example, the amygdala (a triggering site for fear) and the ventromedial prefrontal cortex (a triggering site for compassion) would have to connect to somatosensing regions, areas such as the insular cortex, SII, SI, and the somatosensory association cortices, where body states are continuously processed. Such connections exist [i.e., they have been verified in studies of the neuroanatomy of the brain]. (102)

Fear and compassion are, of course, the constitutive emotions of *catharsis* in Aristotle's classic account. The capacity of the brain to "simulate" a body state in this way is, then, the neurophysiological basis for catharsis, the "triggering" of the feelings of pity and terror in the cortex "as if" we were actually experiencing them in response to real events.

Whether such a simulation results in the purgation of these emotions, as Aristotle stipulated, is another question, and a matter on which the experimental evidence casts some doubt. The psychologist Paul Bloom argues that "catharsis is a poor theory of the emotions, one that has no scientific support. It is just not true that emotional experiences have a purging effect. To take a much-studied case, watching a violent movie doesn't put one in a relaxed and pacifistic state of mind—it arouses the viewer."[75] Experimental evidence suggests that this may be true, but if the spectator is aroused, that is because something like Damasio's as-if body loop is operative, as Aristotle hypothesized was the case in the audience's experience of tragedy. Aristotle may have been wrong about the purging effects of catharsis, but he was right about the embodiment of aesthetic emotions.

When we feel pity and terror in response to a tragic character's fate, that is because the embodiment of emotions can have transpersonal effects. According to Damasio, the "simulation, in the brain's body maps, of a body state that is not actually taking place in the organism" may also make possible the "as if" staging of other people's emotions and feelings. "Because we can depict our own body states, we can more easily simulate the equivalent body states of others" (103, 104). These observations give support to Husserl's distinction between *Körper* and *Leib*, between our lived experience of embodiment and the body as a physical object. "We do not apprehend our body as an external reality among others," the neurophenomenologist Jean-Michel Roy explains, "but as something we are and live."[76] Only if the body is a *Körper* rather than a *Leib* can we simulate its states not only in ourselves but also in others. The lived experience of embodiment allows us to share others' emotions "as if" we were having them too.

Such experiences of embodiment may give rise to a sense of self. Neuroscientifically and phenomenologically, however, the self is not a coherent, unified entity but a process and an event. In a definition that is worded very carefully, Damasio describes the self as "a dynamic collection of integrated neural processes, centered on the representation of the living body, that finds expression in a dynamic collection of integrated mental processes" (9). According to this definition, the self is a paradoxical phenomenon in several related ways. Both "neural" and "mental," it crosses the brain-mind divide without erasing it. A "collection" of processes that are also "integrated," the self is characterized by both multiplicity and synthesis. As a "representation" of the lived body, the self is a sign that both is and is not who we are in the fullness and immediacy of our embodied experience. Finally, and perhaps most importantly, the self is fundamentally "dynamic" because it only "is" through its changes in time.

The neurophenomenologists Gallagher and Zahavi offer a similar view: "The consensus from contemporary neuroscience is that neurological processing is for the most part distributed across various brain regions. There is consequently no real unified neurological center of experience, nor is there any real identity across time that we could label the self."[77] Rather, if the self exists, they argue, its "identity" is our experience of the relatedness of embodied processes of living across time, "the original temporal flow" that is the basis of "the invariant dimension of first-personal givenness in the multitude of changing experiences" (203–4). The self is not "located and hidden in the head" (204), they explain; rather, our sense of self is correlated with and emerges from temporal, synthetic, and multiple neurological processes that are lived and embodied. As Alva Noë dramatically declares, "You are not your brain. The brain, rather, is part of what you are" (7). The sense of self is a phenomenological experience of embodied living in time that has neurological correlates to which it is irreducible, even if processes in the cortex and the brain stem are its biological condition of possibility.

Phenomenologically, the self is nothing more or less than what William James describes as the feeling of "warmth and intimacy" that joins our experiences across the passage of time: "So sure as this present is me, is mine, . . . so sure is anything else that comes with the same warmth and intimacy and immediacy, me and mine. . . . This community of self is what the time-gap cannot break in twain, and is why a present thought, although not ignorant of the time-gap, can still regard itself as continuous with certain chosen portions of the past" (1:239). Hence Merleau-Ponty's claim that "we must understand time as the subject and the subject as time" (422). The durational, horizonal temporality of experience both binds the self to itself and prevents it from ever stabilizing or completely unifying. The temporal, horizonal relatedness of experience is the basis of the sense of self, and this experience is correlated in turn with the way the brain works in time.

Both phenomenologically and neuroscientifically, the temporality underlying selfhood is a foundation that is always slipping away and is never at one with itself. The self and the brain are decentered because they happen in time. Damasio offers an interesting and illuminating although somewhat misleading metaphor to describe these paradoxes. He compares the self to the workings of a symphony:

> Conscious minds result from the smoothly articulated operation of several, often many, brain sites. . . . The ultimate consciousness product occurs *from* those numerous brain sites at the same time and not in one site in particular, much as the performance of a symphonic piece does not come from the work of a single musician or even from a whole section of an orchestra. The oddest thing about the upper reaches of a consciousness performance is the conspicuous absence of a conductor *before* the performance begins, although, as the performance unfolds, a conductor comes into being. For all intents and purposes, a conductor is now leading the orchestra, although the performance has created the conductor—the self—not the other way around. (23–24, emphasis in original)

The symphony metaphor is useful for representing the brain's distributed multiplicity of activities, which are not reducible to a single processing site. Like the metaphor comparing the brain to a cocktail party, the figure of the symphony characterizes the brain on the model of a society rather than an individual, a collection of more or less integrated processes rather than a top-down, unified entity. But the reciprocal, to-and-fro activity of the musicians may suggest better than the cacophony of conversations at a cocktail party the integrating functions of cerebral processing—how the brain "plays," as an orchestra does, and performs various harmonies (and dissonances, and even the occasional wrong note).

The conductor does not preexist the performance, as Damasio cleverly observes, but is produced by it, just as there is no central manager—no man in the machine—directing the brain's activity. Here, however, the metaphor breaks down, or at least demands a different interpretation than Damasio gives it, because the conductor is not the self, any more than James Levine or Seiji Ozawa is (or rather, was) the Boston Symphony Orchestra. The self is the orchestra in and through its performance(s), as it plays over time. The conductor may come to stand metonymically for the orchestra, as a particular, named, objectlike identity may seem to define who someone is, but such a representation is necessarily a misrepresentation of the distributed, temporally fluid, ever-shifting phenomenological and neurological bases of selfhood.

Debates about whether consciousness is unified or divided are often a consequence of splitting apart the paradoxical combination of synthesis and decentered multiplicity that characterizes the self and the brain. For example, Semir Zeki declares that "the statement that there is a single, unified consciousness cannot be true," because the brain consists of multiple "processing sites" that are also simultaneously "perceptual sites" ("Disunity of Consciousness" 214). He prefers to characterize the brain as an ever-shifting, inherently multiple collection of incompletely integrated "micro-consciousnesses." Although John Searle agrees with neuroscience that "consciousness is a biological phenomenon like any other" and "is entirely caused by neurobiological processes" that are "realized in brain structures," he rejects what he calls the "building block approach" and contends that "the brain somehow unites all of the variety of our different stimulus inputs into a single unified conscious experience."[78] He specifically rejects Zeki's notion of a microconsciousness on the grounds that we must first be conscious in order to notice a microconscious state: "I know what it is like for me to experience my current conscious field, but who experiences all the tiny micro-consciousnesses? And what would it be like for each of them to exist separately?" (573).

Searle's spatial metaphor of consciousness as a unified field misrepresents

the temporality of experience, even though his own argument crucially assumes that experience happens in time. This temporality is implicit in a crucial step in his argument: "Only the already conscious subject can have visual
experiences," he contends (574). If we can only notice a microconscious state
after the fact, when an "already" existing consciousness takes aim at it, that is
because consciousness is not "single" and "unified" but happens across time
and is consequently never at one with itself. Hence Merleau-Ponty's assertion that our reflections always disclose a preexisting fund of unreflective
experience that they can never completely catch up with, so that "there is no
thought which embraces all our thought" (xiv). William James offers a typically striking formulation of these temporal displacements: "The effective
consciousness we have of our states is the after-consciousness" (1:644).

Any theory of consciousness needs to account for both the continuity and
the discontinuity of its activities of meaning creation as they occur over time.
Similarly, and correlatively, any theory of the brain must include both its
geographically scattered, temporally disjunctive processes and its capacities
for integration and reciprocal, to-and-fro communication. To argue that consciousness is inherently a disunity, as Zeki does, or a single unified field, as
Searle does, is to split apart the phenomenological and neuroscientific paradoxes of temporal functioning. A both/and rather than an either/or approach
to the relation between multiplicity and synthesis is required by the temporality of consciousness and the decentered brain.

A decentered notion of the self is sometimes thought to be inherently subversive, because it undermines the Cartesian subject's self-certainty ("I think,
therefore I am"). The displacements, disjunctions, and disunities that characterize consciousness are, however, simply a biological fact of life and are
based on the topographical and temporal decenteredness of the brain. Not
intrinsically subversive, negative, or destructive, this decenteredness is con-
ducive to living. (Otherwise it would long ago have given way to the forces of
evolutionary change.) As the neurophenomenologist Jean-Luc Petit explains,
there is an intrinsic instability and decenteredness to life because organisms
are always "on the move": "Before becoming . . . an acting subject, the subject
of its acts, the organism is already 'on the move,' because only as already on
the move is it capable of finding out about itself, discovering its own capacities and mastering them, and indeed of becoming for itself the pole of its own
acts, if not a subject."[79] The inherently unstable, decentered orientation of the
organism toward the "not yet" is what makes it possible for it to be "on the

move" and to act on and respond to its environment in a to-and-fro, exploratory manner.

This reciprocal, recursive, mutually formative process of interacting with the world can take the form of "play." Being on the move in the world can be a way of playing with the world. A brain at one with itself could not play, and it could not be on the move and would not long survive. One of the distinctive practices through which human organisms act on the world in an exploratory manner is reading. A stable, unified brain could not read, because reading is a temporal practice, oriented toward the future. The decenteredness of the brain makes it possible for us to read and in turn makes it possible for reading to shape who we are and who we might become. We can be on the move as readers and enjoy the playfulness of literature because we have decentered brains.

CHAPTER 5

The Social Brain and the Paradox of the Alter Ego

One deficiency of much neuroscience is its neglect of the social dimensions of the brain. This problem is no doubt in part methodological. It is easier, after all, to study an individual brain with an fMRI apparatus or to attach an electrode to a single cell—not that the technology of either procedure is simple—than to map the interactions of a network of brains. As Patricia Churchland notes, "Social life is stunningly complex, as is the brain that supports our social lives."[1] Philosophical assumptions are also certainly to blame, especially the legacy of the Cartesian focus on the consciousness of the single, self-reflecting individual. As the neuroscientist Marco Iacoboni points out, "Philosophical and ideological individualistic positions especially dominant in our Western culture have made us blind to the fundamentally intersubjective nature of our own brains."[2] No brain is an island, however. Evan Thompson reminds us that "stripped of culture, we simply would not have the cognitive capacities that make us human."[3]

We would not, for one thing, be able to read. The neuronal recycling through which education transforms the brain into an apparatus for reading, generation after generation, as a culture of literacy preserves and extends itself, is a prime example of how the neurobiology of our cognitive functions is inextricably intertwined with social interactions that shape and are shaped by them. The evolutionary psychologist Merlin Donald observes that "culture effectively wires up functional subsystems in the brain that would not otherwise exist."[4] This is the case not only with reading but also with a wide range of cognitive, emotional, and even motor functions that develop, or fail to, depending on the contingent, variable interactions of the brain with its environment of significant others, starting with parents but extending to the far reaches of culture and society.

There are three dominant approaches to understanding the social life of the brain, and the emerging consensus is that none alone can account for the complicated, messy work whereby the self and others meaningfully interact. One approach, known as *theory of mind* (ToM) or *theory theory* (TT), focuses on our capacity to attribute mental states to others—to engage in *mind reading*, through which we theorize about the beliefs, desires, and intentions of others, which we recognize may differ from our own. Adherents of this approach regard as a key moment in cognitive development the ability of children at age four or five to pass "false belief" tests, in which they must understand that another child (who doesn't know, for example, that a marble has been moved from box A to box B) will think differently about things than they do (and will look for the marble in A instead of B).[5] Critics of theory theory point out that children have meaningful relations with others before they can understand "false beliefs" (otherwise they couldn't take the tests they fail before age four). "The mind is not a book," Iacoboni argues, and we do not have "to be scientists, like Einstein, analyzing every person around us" with "theories" about their mental states in order to understand "the simple, everyday actions of our fellow humans" (73).

A more plausible view, according to *simulation theory* (ST), is that we interpret others by using our own thoughts and feelings as a model for what others must be experiencing and run "simulation routines," which put us in their shoes. Such imitation can occur right from birth, with more or less conscious awareness, more or less deliberately or automatically, and does not require the ability to "theorize" or "mentalize."[6] Critics of ST object, however, that it begs the question it claims to answer because it assumes that "the simulator already has some idea of what's going on with the other person," which is "the very thing we are trying to explain."[7] In any case, as Christian Keysers and Valeria Gazzola sensibly point out, "social cognitions range from the intuitive examples studied by the simulationists to the reflective ones used by ToM investigators," and an adequate theory of the social brain must account for both.[8]

A third approach may provide the needed bridging mechanism, although there is still some controversy about the experimental evidence underlying it. In the early 1990s a team of neuroscientists in Parma led by Giacomo Rizzolatti discovered *mirror neurons* (MNs) in the motor cortex of the macaque monkey that fired not only when the animal performed a specific action but also when it observed the same action by another monkey or an experimenter.[9] This finding almost immediately led to speculation about whether similar mirroring mechanisms at the neuronal level might underlie such key social behaviors as imitation, learning, and communication. A few years later, Vittorio Gallese, a member of the Parma team, offered the "conjecture . . . that MNs represent a primitive version, or possibly a precursor in phylogeny, of a simulation heuristic that might underlie mind-reading," a proposal he followed up with thoughtful papers about how such a mechanism might work.[10]

The mirror-neuron bandwagon has picked up considerable steam since then, as is evident in the excitement with which the neuroscientist V. S. Ramachandran declares, "These neurons . . . were for all intents and purposes reading the other monkey's mind, figuring out what it was up to. . . . It is as if mirror neurons are nature's own virtual-reality simulations of the intentions of other beings."[11] This enthusiasm has been matched by equal doses of skepticism about whether humans have mirror neurons and, even if they do exist, whether they can bear all the explanatory weight they have been asked to carry.[12] The claim that mirror neurons are, in the words of the UCLA neuroscientist Marco Iacoboni (previously of Parma), "the biological roots of intersubjectivity" (152) is important and intriguing and deserves careful, critical attention, which this chapter attempts to provide (with the caveat that some of the issues have to do with experimental methods and techniques of statistical analysis on which I am not competent to pronounce). It is probably safe to say that the brain's social capacities are much too complex to be explained by a single neuron, but the salutary effect of the mirror-neuron debate is that it has once and for all taken the brain out of the vat and situated it in the social world.[13]

Whatever approach (or, more likely, combination of approaches) eventually emerges from these controversies, the challenge to the neurobiology of social skills is to account for what phenomenology calls "the paradox of the alter ego."[14] The paradox is that these relations are simultaneously both intersubjective and solipsistic—inherently, inextricably, foundationally. As Husserl explains, "I experience the world . . . as an intersubjective world, actually there for everyone. . . . And yet each has his experiences," irreducibly different from others', varying according to the perceiver's particular perspective.[15] As Merleau-Ponty points out, "The social is already there when we come to know or judge it" (362) because the intersubjectivity of experience is primordially given with our perception of a common world. Human existence is characterized by a kind of *primary intersubjectivity* because we exist in a world with other subjects whose different perspectives on the objects we have in common is a guarantee of their independence and completeness even if we only have these objects partially, in aspects, with sides and dimensions beyond our horizons.[16] And yet, Merleau-Ponty goes on, "there is . . . a solipsism rooted in living experience and quite insurmountable" (358) for the very reason that the ego can never have the self-experience of the alter ego without collapsing the difference between them. I am destined never to experience the presence of another person to him- or herself, and my self-for-others is also necessarily a mystery to me.

Husserl consequently asks: "[Are] these two primordial spheres not separated by an abyss I cannot actually cross, since crossing it would mean, after all, that I acquired an original . . . experience of someone else?" (121). Behind this question lurks a paradox, however, because, in Merleau-Ponty's words, "my experience must in some way present me with other people, since otherwise I should have no occasion to speak of solitude, and could not begin to pronounce other people inaccessible" (359). Solipsism is a problem only for an intersubjective self who already perceives the other as a coequal subject in a world they share.

These paradoxes are evident in some of the contradictions that afflict theory theory and simulation theory, contradictions that only seem like problems until we recognize their roots in the paradox of the alter ego, at which point they become evidence of the enigmatic combination of intersubjectivity and solipsism that characterizes social relations and must have correlates in the neurobiology of our social skills. There is a need for a theory of mind and mind-reading skills only because of the unsurpassable solipsism that makes it impossible for me to experience another's self-presence. But the accusation that this ability is secondary to a prior understanding of the other on which it builds is in turn evidence of our primary intersubjectivity, an original sense of the other's commonality with me, without which I would have no basis for theorizing about his or her inner states. Simulation theory assumes that we find in the alter ego an analogue of ourselves, but this is paradoxical and mysterious in ways its critics point out only because ego and alter ego are intrinsically, simultaneously, both similar and different. As Husserl observes: "The other is a 'mirroring' of my own self and yet not a mirroring proper, an analogue of my own self and yet again not an analogue in the usual sense" (94). How to capture this mysterious, paradoxical, and yet lived, ordinary, and everyday experience of the alterity of an ego who is both "like" and as such "not-like" me (because "like" is not "same")—that is the challenge for an adequate account of the relation of self and other.

Such an account must recognize that mirroring and analogizing are duplicative acts—doublings of "me" and "not-me," "like" and "not-like"—that human beings routinely, automatically, and unreflectively engage in as they negotiate their way in a paradoxically intersubjective and solipsistic world. As Husserl explains, "Ego and alter ego are always and necessarily given in an original 'pairing'" (112). This pairing of me and not-me makes mirroring and analogizing paradoxical processes of doubling—"das Gleich-setzen des Nicht-Gleichen," to recall Nietzsche's useful phrase, the setting equivalent of what is (as an equivalence) not identical.[17] The capacity of human beings to engage in doubling of this kind is based on bodily, neuronal, and cortical functions that constitute the neurobiological foundation of our social skills. Processes of doubling underlie theory theory, simulation theory, and the activity of mirror neurons.

Reading also is an experience of doubling. The paradox of reading, according to Georges Poulet's classic description, is that I think the thoughts of another, but I think them as if they were my very own: "Because of the strange invasion of my person by the thoughts of another, I am a self who is granted the experience of thinking thoughts foreign to him. I am the subject of thoughts other than my own. My consciousness behaves as though it were the consciousness of another."[18] Reading is a peculiarly solipsistic and intersubjective experience that simultaneously enacts and overcomes the paradox of the alter ego. As Iser points out, reading is on the one hand an experience of solitude and isolation in which one does not immediately encounter one's interlocutor: "with reading there is no face-to-face situation."[19] And yet, on the other hand, this private experience allows us to feel and know from the inside the presence of others to themselves and to see the world through the eyes of others as we cannot in real life. The grand illusion of literary art, according to Henry James, is that it "makes it appear to us for the time that we have lived another life—that we have had a miraculous enlargement of experience."[20] This other world exists, however, only because we bring it to life through our own acts. As Jean-Paul Sartre observes, "The literary object has no other substance than the reader's subjectivity; Raskolnikov's waiting is my waiting which I lend to him. . . . His hatred of the police magistrate who questions him is my hatred which has been solicited and wheedled out of me by signs."[21] If we identify with characters like Dostoevsky's anguished protagonist in *Crime and Punishment*, this is not simply a transcendence of the bound-aries between self and other but a doubling of me and not-me, through which my acts paradoxically involve me intimately in another world that both is and is not my own and that I animate with my own powers of meaning creation.

Reading entails a paradoxical duplication of consciousness. As Iser explains, "Every work we read carves a different boundary within our personality": "In thinking the thoughts of another, [the reader's] own individuality temporarily recedes into the background, since it is supplanted by these alien thoughts, which now become the theme on which his [or her] attention is focused. As we read, there occurs an artificial division in our personality, because we take as a theme for ourselves something that we are not."[22] Our self does not disappear, since only by its exercising its powers does the other world emerge. Instead, according to Iser, reading produces a duplication of selves, an interplay of the "alien me," whose thoughts I re-create and inhabit, and "the real, virtual 'me,'" whose horizons are temporarily changed by the experience (hence the sense of "strangeness" of the "invasion" of our consciousness by the consciousness of another that Poulet reports).

This doubling can take different forms, and have different consequences, with different literary works and different readers. It may make possible an immersion in another world whose illusion carries us away for a time from our customary sense of things, or it may confront us with alien ways of thinking, feeling, and perceiving whose foreignness calls into question characteristics of our customary understanding of the world that we had perhaps not noticed until their boundaries were exposed by this juxtaposition. Reading can be such a variable experience with different texts and for different readers because it can enact the paradoxical doubling of self and other in an unpredictable variety of ways.

This makes reading an excellent experience through which to study the paradox of the alter ego. Asking about the brain's capacity to read can provide insight into its social skills. The neurobiological explanations of our ability to understand others should make sense of the paradoxes of reading, and the paradoxes of reading are in turn a good test of the claims made by competing theories of the social brain. What does the experience of reading reveal about the neurobiology of intersubjectivity, and what does the brain's capacity to "double" self and other tell us about how we read?

A process of doubling, or pairing, is evident in the workings of mirror neurons. The initial experiments showed that about 20 percent of the neurons in

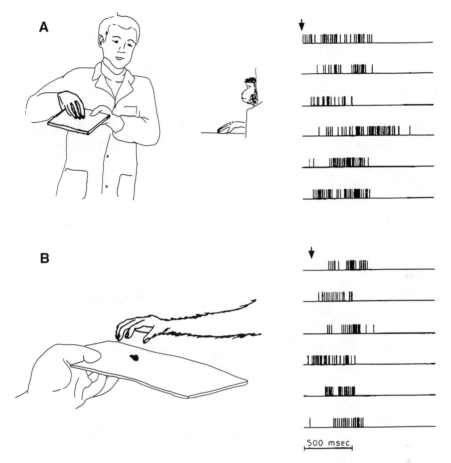

Figure 5.1. Visual and motor responses of a mirror neuron. Comparison of the electrical signals recorded from a mirror neuron when observing an action (*A*) and when performing the same action (*B*). Reproduced by permission from G. di Pellegrino et al., "Understanding Motor Events: A Neurophysiological Study," *Experimental Brain Research* 91 (1992): 178 (fig. 2).

the motor cortex of the macaque monkey fired not only when the animal performed an action like grasping a piece of food but also when it observed the same action (see fig. 5.1).[23] Further experiments indicated that these neurons were responding not to the general shapes of motion but to particular, goal-directed actions. Some neurons were *strictly congruent*, responding only to a very specific action (twisting a raisin in one direction but not another). Other mirror neurons were *broadly congruent*, reacting with equal strength to actions within a range defining a particular motor category (grasping with the whole hand, for example, as well as with a precision grip). Some mirror neurons were *multimodal*, responding to sounds as well as to visual cues (hearing or seeing a peanut being broken). Further, the monkey's mirror neurons only fired in response to actions within its motor repertoire, not, for example, to an action performed by a tool, although it reacted when a human or animal hand performed the act (macaque monkeys don't use tools). Similarly, the monkey's mirror neurons were indifferent to mimicked actions (macaques don't pantomime) that they responded to when they were actually performed.

These findings led Rizzolatti to propose a theory of "action understanding," which posits a primary, intuitive relation between self and other: "mirror neurons allow our brain to match the movements we observe to the movements we ourselves can perform, and so to appreciate their meaning" (Rizzolatti and Sinigaglia xii). This "direct-matching hypothesis," as he calls it, claims that action understanding occurs "immediately on the basis of our motor competencies alone, without the need of any kind of reasoning" (xii): "an action is understood when its observation causes the motor system of the observer to 'resonate.'"[24] Instead of actions being the object of interpretive processes that indirectly construe their meaning, the "resonance" between observation and action provides an immediate connection between self and other. Not requiring a theory or a simulation, this resonance is bodily, unreflective, and automatic. It is the biological equivalent of primary intersubjectivity.[25]

As the paradoxical phrase *direct matching* suggests, however, this is not an identity but a pairing (a "match" can only occur, after all, between two entities that aren't the same). The difference between self and other is replicated— only a subset of the monkey's cortical motor neurons fire, which is not what would happen if it were acting—even as it is overcome. This difference is also evident in the very generality of the mirror neurons' responsiveness. The distinction between *strict congruence* and *broad congruence* suggests a range of recognition rather than a simple identity of action and observation. That the resonance only happens with actions within the observer's motor repertoire (and not with tools or pantomime for the macaque monkey) is further evidence of the self-other difference that the mirroring process can sometimes (but not always, and never completely) transcend. Mirror neurons create an immediate bond, then, that is evidence of primary intersubjectivity, even as the very process of mirroring and matching entails a doubling of ego and alter ego.

Motor mirror neurons apparently are sensitive not only to actions but also to the intentions underlying them and the goals at which they aim. In one experiment, the responses of mirror neurons were measured when monkeys grasped a piece of food in order to eat it as opposed to when they picked up food and placed it in a container on their shoulder (so that the motion would be nearly equivalent to eating).[26] Approximately 30 percent of a monkey's mirror neurons fired in response to either action, but the rest differentiated between grasping to eat and grasping to place even though the object and the type of action were almost identical. (Perhaps not surprisingly, 75 percent of the neurons were more interested in eating than in placing.) Without theorizing or simulating, it would seem that these neurons were intuiting another's intentions. This inference is reinforced by another experiment, in which monkeys' view of the goal of an action was blocked although they were shown what was behind the occlusion. When there was an object behind the screen, about half of the neurons fired that had also responded when the view wasn't blocked, but none fired at all when the monkeys knew there was nothing there, even though the motions transpiring before their eyes were identical (see fig. 5.2).[27] The neurons that first fired and then didn't seemed to interpret the same motions as different actions when the goals differed. In both experiments, the monkey's mirror neurons seemed to be responding not simply to a physical state of affairs visible to observation but to distinctions having to do with another agent's aims and purposes. If this is so, these neurons truly are the glue that binds primary intersubjectivity.

Some of the most important criticisms of the mirror-neuron theory are directed at the supposition that these neurons can "read" goals and intentions. According to Gregory Hickok, the ability to infer meaning automatically from behavior assumes that "self-generated actions have an inherent semantics and that observing the same action in others affords access to this action semantics." He points out, however, that "the motor act of pouring liquid from a bottle into a glass could be understood as *pouring, filling, emptying, tipping, rotating, inverting, spilling* (if the liquid missed its mark), *defying/ignoring/rebelling* (if the pourer was instructed not to pour), and so on. A motor representation cannot distinguish between the range of possible meanings associated with such an action" (1231, 1240). Patricia Churchland voices similar doubts: "I may raise my arm, motivated by any number of completely different intentions: to ask the teacher a question or signal the soldiers to charge or reveal my position to my hunting group or to stretch out

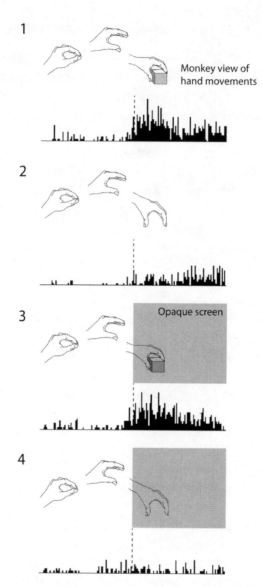

Figure 5.2. Mirror-neuron responses in the macaque monkey to occluded objects. A mirror neuron fires when the monkey views a hand grasping an object (1) but not when the same motion is performed without an object (2). This neuron still fires when the monkey cannot see the object behind a screen but knows an object is there (3), but not when it knows no object is present (4). Reproduced by permission from M. A. Umiltà et al., "I Know What You are Doing: A Neurophysiological Study," *Neuron* 31 (2001): 158 (fig. 1).

my shoulder muscles or to vote for building a school, and so on and on. . . . Merely mirroring a movement will not tap into the range of higher-order intentions, or select the right one, for which a lot of background understanding, probably including a theory of mind, is needed" (140).[28] How often have I asked a student if her raised hand meant she had a comment to offer only to be told that she was just scratching her head? As Churchland somewhat acerbically notes, "A neuron, though computationally complex, is just a neuron. It is not an intelligent homunculus" (142).

No single neuron will provide the solution to the problem of other minds. These objections are a reminder that relations with others extend along a broad continuum, from immediately intuitive understanding to the befuddlement caused by ambiguous or opaque behavior. Even Rizzolatti, discoverer and advocate of mirror neurons, acknowledges the existence of fMRI evidence showing that "other more cognitive ways of 'reading minds'" involving nonmotor areas of the cortex may be invoked in anomalous cases where immediate understanding breaks down (his examples are "a motor act in an implausible context" and the need "to judge whether the intention of the observed action was ordinary or unusual").[29]

The immediate, unproblematic understanding that I have of another's actions most of the time may indeed be unthinking and automatic because it is based on resonances between my motor system and the actions I observe. Our unreflective sense of sharing a common world sometimes falters, however, just as some actions may have multiple meanings, and here theorizing may be required. We ordinarily don't need to hypothesize self-reflectively about someone else's actions and intentions, but we sometimes do, especially when immediate understanding fails. And then all of the brain's hermeneutic capacities may come into play—its ability to fit evidence into patterns, to project hypotheses about hidden sides, to imagine perspectives consistent with but horizonal to what is open to our view—and these processes involve reciprocal, to-and-fro interactions among a variety of cortical regions, including but not limited to the areas where mirror neurons can be found.

One challenge to the mirror-neuron theory is how to explain our ability to understand actions we have never performed or witnessed. Churchland's examples (mine would be different, as would yours) are "skinning a rabbit, skidding logs, or traveling on a zipline or a hang glider" (141). Perhaps less exotically, Hickok points out that "musically untrained people can recognize, say, saxophone playing even if they have never touched the instrument, just as one can recognize actions of non-conspecifics (barking, flying)" that lie outside our motor capacities (1236). But the saxophone example also suggests the potential usefulness of motor knowledge to understanding an action we cannot perform, as Hickok himself concedes: "Knowledge of how to grasp a saxophone, finger the keys, and position one's mouth on the mouthpiece, can . . . augment the abstract concept" we may have of saxophone music "by providing a specific sensory-motor association" and may consequently enhance our understanding of it (1240). Knowing something about how a violin works, for example, certainly increases our understanding and appreciation of orchestral music, and learning to play even one instrument may improve our ability to listen with discrimination to other instruments (the trumpet is not a major component of an orchestra, but my experiences in marching band as a child still help me listen to symphonies and even operas with little brass).

These examples all show how, applying the hermeneutic circle, we extend what we know to make sense of unfamiliar phenomena. If observing someone skinning a rabbit does not trigger mirror neurons from similar actions in our motor repertoire, we may not have an immediate intuitive understanding of what the hunter is up to, but we can interpret it nevertheless by extrapolating from similar actions we are familiar with, with similar goals and intentions, thereby using the hermeneutic resources that simulation theory and theory of mind foreground. Doing so, however, depends on an assumption of primary intersubjectivity, which undergirds the supposition that another acts as I would if I were over there doing what he or she is doing. The experience of primary intersubjectivity is something motor mirror neurons provide, even if the resonance between action and observation alone cannot always by itself explain the meaning of others' behavior.

Reading literary texts can set in motion all of these ways in which self and other interact, from the primary intuitive resonance of mirroring to various levels of simulation and explicit theorizing about motives, goals, and intentions (including meta-theorizing about the interpretive processes themselves through which we seek to know what others are thinking). Iser points out that "the dynamic lifelikeness" of a represented world "enables us to absorb an unfamiliar experience into our personal world" because we immerse ourselves in illusions that we ourselves create and identify ourselves with (*Implied Reader* 288). He quotes a report from one of Charlotte Brontë's early readers: "We took up *Jane Eyre* one winter's evening, somewhat piqued at the extravagant commendations we had heard, and sternly resolved to be as critical as Croker. But as we read on we forgot both commendations and criticism, identified ourselves with Jane in all her troubles, and finally married Mr. Rochester about four in the morning."[30] As this reader wryly implies, such identification is a doubling of me and not-me (they both did and did not marry Rochester, after all), but the power of an immersive experience like this to carry a reader away depends on intuitive resonances of a bodily, immediate, and unreflective kind that disarm critical distance.

Even with texts that encourage such immersion, however, experiences of primary mirroring can alternate with critical, cognitive reflections, especially when the illusions that prompt involvement are interrupted or disturbed by alien associations that suggest different patterns of consistency building. "As we read," according to Iser, "we oscillate to a greater or lesser degree between the building and the breaking of illusions" (288)—between involvement in the gestalts we create and observation of their contours. Such oscillation may characterize any reading experience and may vary according to the patterns of the text or the proclivities of the reader. Some writers may be particularly intent on disrupting consistency and preventing the formation of illusions in order to prompt explicit reflection about hermeneutic processes that immersive texts exploit less self-consciously. (Iser cites Beckett and Joyce, but the list could be expanded to include Conrad, Faulkner, Pynchon, Robbe-Grillet, and many other experimental modernist and postmodern writers who seek to write in ways that hew more closely to the manner in which we know the world but, as a result, may create puzzling, if intriguing and sometimes fantastic texts.) Other interruptions may aim to enhance consistency building and identification, as when Balzac's omniscient narrator intervenes to inform the reader that a particular character, setting, or practice is emblematic of the "laws of Paris," or when George Eliot's wise, sympathetic, but also critical narrator suggests that we should temper our judgments of a character's follies with a recognition of moral vulnerabilities we all share. Illusions may be temporarily suspended by such reflections, but this kind of interruption seeks to buttress and reinforce verisimilitude by providing patterns through which we can theorize or simulate the workings of other minds.

We can have a sense of living in another, quasi-real world when we read only because we employ analogous processes to knowing others in life. These processes range, in reading and in life, from the unreflective pairing of mirror resonance to different degrees of self-conscious simulating and theorizing.

Sometimes our involvement with a character or with an author's world is as immediate and intuitive as Wittgenstein claims our everyday knowledge of another's state can be: "We *see* emotion. . . . We do not see facial contortions and *make the inference* that he is feeling joy, grief, boredom. We describe a face immediately as sad, radiant, bored, even when we are unable to give any other description of the features."[31] This observation is supported by experimental evidence that seeing another's facial expression activates mirroring processes of emotion recognition.[32] Similarly, we know that Anna Karenina is desperately unhappy before her suicide without having to mentalize. But sometimes characters and texts are anomalous and puzzling. Alvin Goldman distinguishes between "low-level simulation," which involves mirror processes, and "high-level simulation," which employs mental "pretense" or what he calls "enactment imagination" in adopting the perspective of someone else.[33] Why Anna chooses suicide rather than another course of action and whether Tolstoy is unfairly driven by ideological considerations in killing her off (as D. H. Lawrence famously criticized him for doing)—such questions may require high-level acts of simulation to re-create the emotional, psychological, and even moral frame of mind of a character or an author whose perspectives might seem foreign or unusual.

Much of my work as a teacher of narrative is devoted to helping students learn how to undertake such simulations. Blakey Vermeule has this value of literature in mind when she asks, "Why do we care about literary characters?" and answers, with good reason, "gossip: we need to know what other people are like, not in the aggregate, but in the particular."[34] She explains that although gossip has a bad name, it is not necessarily a bad thing inasmuch as "we scan other people because we have to cooperate with them and compete against them," and gossip may help (among other things) to identify whom we can and can't trust (33). Gossip can be the "idle talk" (*Gerede*) that Heidegger rightly disparages when it is motivated by uncaring, unconcerned "curiosity" (*Neugier*), which seeks entertainment without engagement and pretends to know what it doesn't really understand.[35] But one of the important ways in which literature is useful for life is that reading novels, plays, and poems and discussing them with other readers provides a laboratory in which the brain can experiment with its social skills—testing, challenging, extending, and scrutinizing habitual practices for knowing others that may go unnoticed or seem simply "natural" in our everyday engagement with other people. Texts (from earlier periods, say, or other cultures) that dramatize particularly unfamiliar ways of thinking, feeling, and judging are especially useful for these purposes because they defamiliarize and thereby expose for scrutiny what we take for granted not only at the level of primary intersubjectivity but also in our more explicit simulating and theorizing habits.

Not simply a matter of representational content, this testing can happen in important ways in the lived experience of reading itself. Literary form matters because, among other reasons, the strategies and devices through which texts manipulate our identifications with characters set in motion— and thereby play with and often surprise—the everyday habits and assumptions through which we deal with the social world. As we read, these practices are made available for scrutiny with special immediacy and potential effect because they are not something we merely observe in others whose lives a text dramatizes but are also directly and experientially "there" in our own relation to the text's world. As we build illusions that allow us to immerse ourselves in a represented world, the text's forms play with these involvements and sometimes pull us up short, breaking what we ourselves have made and are therefore immediately implicated in and concerned about. Such movement back and forth between immersion and reflection distinguishes literary "gossip" from everyday "idle talk" by giving the simulating acts of literary reading a dimension of engagement and self-criticism that is lacking in our nosy, indifferent curiosity about a neighbor's messy love life or financial woes. Different textual strategies for staging, encouraging, and resisting our identifications with literary characters make our experiences as readers part of the experiment, testing our habits for understanding other worlds, even as we also occupy the role of the experimenter, who learns from what this laboratory animal (ourselves) undergoes. If we can live another life by reading, as Henry James tells us, then examining the very process of reading can itself be a means of analyzing how our lives interact with other lives in the everyday social world.

Many of the strategies and devices that narratologists study are ways of staging the relation of self and other in the reading experience.[36] The choice of a narrator, for example (first- or third-person, reliable or unreliable), will set up a different relation in the reader's consciousness between the real me and the alien me, through which the paradox of the alter ego manifests itself as we read. The "narratee" whom the narrator addresses both is and is not the reader, and this doubleness makes unreliable narration possible as the reader questions whether the narrator is attempting to pull the wool over the narratee's (that is, our) eyes. One reason why questionably reliable narrators are so often examined by literary theorists interested in epistemological questions is that these doubled structures open for scrutiny and self-conscious reflection relations of doubling that otherwise operate invisibly in our experience of other people and narrative worlds.

The question of how a story is "focalized"—the point of view through which events are seen—similarly calls attention to the doubling of the reader's consciousness with the perspectives we occupy in the text. Experiments with focalization can vary this distance for different purposes, as when, for example, indirect discourse makes it difficult to disentangle the narrator's viewpoint from that of a character whose thoughts are being rendered and to decide how to calibrate our sympathies and judgments.[37] Stephen Dedalus is clearly the author's alter ego in Joyce's *Portrait of the Artist as a Young Man*, for example, but inhabiting his perspective is an uncannily unstable experience because we are often unsure whether to sympathize with him or to view his deficiencies with an irony and criticism he lacks, even as we are inside his head. Such narrative devices can have many functions, but all of them depend on the doubling between the real me and the alien me in reading through which narrative re-creates in an "as if" manner our experiences with other worlds.

The simulations we engage in as we read or otherwise seek to know others may entail theorizing, but not necessarily. Citing fMRI evidence, Grit Hein and Tania Singer point out that affective empathy and "cognitive perspective taking" set in motion "different neural networks" (the insula and the anterior cingulate cortex [ACC] for emotional identification as opposed to various prefrontal, temporal, and temporal-parietal lobes for theoretical mind reading).[38] If we simulate the emotional state dramatized in a character or staged by a text like a lyric poem, we will in all probability activate different cortical areas than we would in theorizing about why such feelings came about and what they imply. A brain scan of someone crying at Little Nell's death in *The Old Curiosity Shop* will certainly light up different areas than a scan of someone tracing the complexities of point of view in *Lord Jim* or *The Golden Bowl*. Acknowledging that "experiencing the same emotion as another is not sufficient to infer the cause of that emotion and hence is only a first step for mentalizing," the TT advocates Chris and Uta Frith identify various brain regions (mostly in the frontal cortex) that are the "neural correlates of mentalizing"—different, particularized cortical bases for such acts as "perspective taking," application of "knowledge of the world" in reading context, and "anticipating what a person is going to think and feel and thereby predict[ing] what they are going to do."[39] All reading passes through the brain's letterbox, but imaging experiments would be expected to show different areas of the cortex lighting up for reading experiences that were either more intuitive or more reflective, with the insula and the ACC more active in response to texts encouraging emotional involvement and immersion, for example, whereas various areas in the frontal, temporal, and parietal areas would show more activity in response to texts that call for the kinds of mentalizing in the Friths' catalog.

Which areas of the brain get activated, in which combination, in a particular reading experience is no doubt fluid and complex and cannot in all likelihood be pinpointed with mechanical, deterministic precision. As a recent review article sensibly warns, "We should . . . be wary of treating different brain regions as separate entities. Simulation is a highly integrated process which is likely to depend on the networks connecting various regions."[40] Theorizing about a character, an author, or a text is also highly integrated and likely to involve to-and-fro interactions among a variety of neural networks. Even texts that emphasize one pole or the other—immersive identification as in *Jane Eyre* or the sort of self-conscious mind reading that Henry James notoriously dramatizes in late works like *The Ambassadors* or *The Wings of the Dove*, not to mention what happens when we read Joyce's experiments with style in *Ulysses*—are multidimensional in the cortical processes they invoke, from mirroring to emotional simulation to cognitive perspective taking. It would be wrong to reduce reading to any one of them. (And then there are readers like Oscar Wilde, who read "against the grain," as when he notoriously declares that he can't help laughing at Nell's sentimental death throes—one can only imagine what his brain scan might have looked like.)

These complications suggest why Lisa Zunshine's explanation of "why we read fiction"—"because it offers a pleasurable and intensive workout for my Theory of Mind"—is at best only partially correct.[41] Brian Boyd criticized Zunshine for, among other things, ignoring how we also get pleasure from simple fictions that don't push our ToM beyond our "comfort zone," to which Zunshine replied that "fictional narratives endlessly *experiment with* rather than *automatically execute* our evolved cognitive adaptations."[42] Literary works do both, however, in differing degrees, and can provide pleasure in doing so. The phenomenology of reading is a many and various thing. To view the experience of reading as a variable process of illusion making and illusion breaking, immersion and reflection, consistency building and the disruption of pattern, is an attempt to offer a model flexible and capacious enough to do justice to the full range of aesthetic phenomena that different readers have discovered and described. The neural processes correlated to this phenomenology will necessarily be similarly diverse and variable.

Before analyzing these processes in more detail, we should also briefly consider questions that have been raised about whether humans have mirror neurons at all. These questions arise because the invasive techniques of single-cell measurement, which the Parma group employed with macaque monkeys, are not ordinarily permissible with human subjects. In an oft-cited study, Ilan Dinstein objects that fMRI evidence of mirroring processes in humans cannot definitively show that the same cells are firing during both action and perception, because too many neurons are covered by the voxel, that is, the smallest unit a scan reproduces.[43] A large number of PET and fMRI studies have shown coactivation of action and observation in areas of the human brain analogous to the macaque monkey's mirror-neuron system—but some have not. Christian Keysers summarizes: "For each experiment that fails to find evidence for mirror neurons in humans there is at least one that succeeds," an ambiguity that can be seen either to support or to question the MN hypothesis, depending on which side of the controversy you happen to back.[44] Recently, however, ingeniously (and ethically) bypassing the prohibition on single-cell experiments with human subjects, wires inserted into the brains of epilepsy patients in order to identify seizure-prone areas for surgical removal were used to search for human mirror neurons, and the results were unambiguously positive. These probes "recorded activity from 11 neurons that behaved exactly like broadly congruent mirror neurons in the monkey: they discharged during both the observation and execution of one type of action."[45] Although it will take time for this finding to be replicated and then assimilated by the neuroscientific community, it seems to be conclusive evidence that humans have mirror neurons.

If they do, this would confirm massive indirect evidence for the existence of mirroring cells in the human brain.[46] The earliest such evidence was discovered in 1954, when Henri Gastaut found that "mu waves" measured by EEG technology desynchronized not only when an action was performed but also when it was observed. After Rizzolatti's group reported the existence of mirror neurons in monkeys, V. S. Ramachandran repeated Gastaut's experiment and proposed at a meeting of the Society for Neuroscience in 1998 that mu-wave suppression was caused by human mirror neurons (123–24). These experimental results are consistent with widespread reports of mirroring with various kinds of actions as well as with pain, disgust, and other emotions, like fear.[47] For example, Iacoboni describes an fMRI experiment that compared "the brain activity of two groups of dancers watching videos" and "found that classical ballet dancers had higher mirror neuron activity than capoeira experts while watching classical ballet videos" and that the opposite was the case with films of capoeira.[48] What is more, the brains of the male and female ballet dancers responded more when they watched moves specific to their gender (lifting versus dancing *en pointe*).[49] Although much detailed and complicated technical work remains to be done to identify and describe the underlying neurobiological mechanisms involved, the balance seems to have tilted toward the recognition that, as Ramachandran puts it, "small groups of cells with mirror-neuron-like properties are found in many parts of the [human] brain"; he cautions, however, that "we must be careful not to attribute all puzzling aspects about the brain to mirror neurons. They don't do everything!" (145).

One especially intriguing property of some mirror neurons is that they respond to the observation not only of actions but also of objects on which these actions have been and can be performed. What are called *canonical neurons* fire, for example, both when a monkey observes an experimenter grasping a cup and when it simply sees the cup. According to Rizzolatti, this is the neurobiological basis of what psychologists call *affordance*, whereby the perception of an object selects the "properties that facilitate our interaction with it" and focuses on "the practical opportunities that the object offers to the organism which perceives it."[50] For canonical neurons, Rizzolatti observes, "the sight of the cup is just a preliminary form of action, a call to arms so to speak," and "objects are simply hypotheses of action" (Rizzolatti and Sinigaglia 49, 77). As David Freedberg and Vittorio Gallese explain, with these neurons "the observation of a graspable object leads to the simulation of the motor act that the object affords," and this mirroring is not unique to monkeys: "brain imaging experiments in humans have shown that observation of manipulable objects like tools, fruits, vegetables, clothes, and even sexual organs leads to the activation" of cortical areas "involved in the control of action" relevant to those objects.[51]

Canonical neurons fire because an object represents past action and holds ready possibilities for future action. In some cases (fruits, vegetables, and sexual organs) these possibilities exist simply as a consequence of the intrinsic properties of the object and may or may not involve us with other agents (although the last item on the list very probably if not necessarily would). But in other cases (tools, clothes, as well as other human artifacts) these potentialities are a product of a history of action that has left its traces in the object's affordances. In such instances, canonical neurons respond to the traces of agency embedded in objects and engage us with other agents in a web of intersubjective relations.

Canonical mirror neurons are no doubt involved in our response to cultural objects of all kinds. The affordances that trigger them recall Heidegger's notion of tools and other kinds of equipment that are "ready-to-hand" (*zuhanden*) because they seem preprogrammed for our active engagement. As Dan Zahavi explains, "It is . . . a fundamental feature of such entities that they all contain references to other persons."[52] The peasant shoes that Heidegger famously reflects on in his essay "The Origin of the Work of Art" can suggest a whole world of activities, involvements, and concerns because our canonical neurons resonate with the agency they imply.[53] These resonances explain why, as Merleau-Ponty observes, "in the cultural object I feel the close presence of others beneath a veil of anonymity" (348). This is not mystical hocuspocus but evidence of the embodiment of primary intersubjectivity: "the very first of all cultural objects, and the one by which all the rest exist, is the body of the other person as the vehicle of a form of behavior" (348). In responding to the affordances in cultural objects, canonical neurons resonate to a kind of embodiment of agency.

Because of the workings of canonical neurons, we may experience indirect but nevertheless bodily resonance with others through a whole range of artifacts, from tools to works of art (including books), that are part of the human motor repertoire. Cultural objects enact the paradox of the alter ego because this resonance puts us in immediate but also mediated relation with the agency of others who are both "there" and "not-there." The potentiality of interacting with others through objects that give evidence of their agency makes others present in these artifacts, even if this presence is only "veiled" and "anonymous" (to recall Merleau-Ponty's words) because they are also absent. This absent presence is what causes canonical neurons to fire.

Although it is important not to get carried away by speculation, the role of canonical mirror neurons in aesthetic experience unquestionably deserves further inquiry. Freedberg and Gallese note that experimental research has shown that "the observation even of static images of actions leads to action simulation in the brain of the observer," and they conclude that "it stands to reason that a similar motor simulation process can be induced by the observation of still images of action in works of art. It is not surprising that felt physical responses to works of art are so often located in the part of the body that is shown to be engaged in purposive physical actions, and that one might feel that one is copying the gestures and movements of the images one sees" (200). This is a hypothesis, not a firm finding, but one worth testing. Although Cinzia Di Dio and Gallese acknowledge that the noisy, claustrophobic confines of an fMRI machine hardly duplicate the conditions of a museum, they report that viewing images of classical sculptures invoked "motor resonance congruent with the implied movements portrayed in the sculptures."[54] Based on the canonical-neuron research, Freedberg and Gallese speculate that "even a still-life can be 'animated' by the embodied simulation it evokes in the observer's brain" (201).

Further, noting experiments showing that "the observation of a static graphic sign evokes a motor simulation of the gesture that is required to produce it," Freedberg and Gallese "predict that similar results will be obtained using, as stimuli, art works that are characterized by the particular gestural traces of the artist," as in Jackson Pollock's extremely dynamic drip paintings (202). If "our brains can reconstruct actions by merely observing the static graphic outcome of an agent's past action," they reason, then "the observation of the gestural traces of the artist" should trigger embodied simulation (202, 201). Canonical mirror neurons would respond not only to content, then, but also to form—not only to representations of actions or manipulable objects but also to the actions that produced the image.

If language is a mode of action that enables simulative experiences, then mirror neurons may also be involved in processes (like reading) in which representation is linguistically mediated. Although the evidence is somewhat controversial and unsettled, MN advocates argue that Broca's area, one of the key language regions of the human brain, is the anatomical equivalent of the F5 premotor cortex in the macaque monkey, where most mirror-neuron activity in the monkey occurs (see fig. 5.3).[55] When directed at Broca's area, *transcranial magnetic stimulation*, or TMS, which can temporarily disable par-

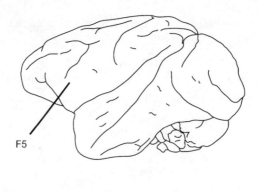

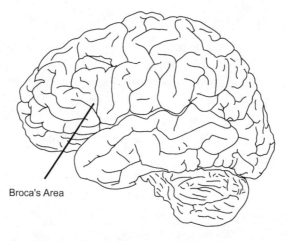

Figure 5.3. Monkey and human brain mirror-neuron areas. Area F5 of the macaque monkey brain is regarded by mirror-neuron scientists as anatomically analogous to Broca's area in the human brain. Drawing by Maggie Buck Armstrong.

ticular cortical regions with pinpoint accuracy and thereby simulate the effects of a brain lesion, has been shown to impair not only speech but also the imitation of elementary motor actions like finger movements.[56]

This finding is consistent with other experimental evidence that links language to embodied imitation. Iacoboni reports that "while listening to other people talk, listeners mirror the speaker with their tongues," an observation confirmed by an fMRI study showing that "the same speech motor area that was activated during speaking was also activated during listening" (104).

Analogously, he notes, when "TMS pulses knocked down the subjects' motor speech areas, their ability to perceive speech sounds was also reduced" (105). Other experimental studies have shown specific links between particular motor areas and the perception of action verbs associated with them. When we read the verbs *throwing* or *kicking*, for example, the cortical areas associated with those actions fire.[57] Iacoboni has conducted similar experiments showing that "human mirror neuron areas for hand movements and for mouth movements . . . were also selectively activated while subjects were reading sentences describing hand actions and mouth actions" like "grasp the banana" or "bite the peach" (94).

These findings have renewed interest in the theory that language has roots in the body and that speech is based on gesture. "Gestures lead; speech follows," according to Iacoboni (86). Long before these technologies were invented, Merleau-Ponty argued that "the body is a power of natural expression" and "the spoken word is a genuine gesture" that "contains its meaning in the same way as the gesture contains its," with a kind of immanence that can resonate bodily. "The gesture does not make me think of anger, it is anger itself" (181, 183, 184). There is more to language than speech, of course, as the Chomskyites and other structuralists are quick to point out, and motor action or gesture cannot account for all of the grammatical, logical, and syntactical dimensions of language.[58] Nevertheless, the findings linking motor processes and language suggest how linguistic activities, like writing and reading, may connect self and other in a bodily, if indirect, way.

Such mirroring is evident in the intersubjectivity of sentence meaning. Sentences are manipulable artifacts that bear traces of human agency and that may trigger our canonical mirror neurons in response. Roman Ingarden argues, for example, that the meaning units in sentences are characterized by "derived intentionality," "a borrowed intentionality, one that is conferred on them by acts of consciousness" that they are no longer in direct contact with (the writer is not present and may even be dead) but that nevertheless still somehow inhabit them. "Because of this detachment from concrete acts of consciousness," with all of their "primary vividness and richness, the intentional correlates of meaning units also undergo various other modifications," including "a certain schematization of their content."[59] In order to be once again animated and filled out, this derived intentionality, originated by the meaning-creating activity of the writer, needs the activity of the reader. Its "schematization" is evidence that it is only secondary, not immediate and original, intentionality. But the paradox of linguistic representation is that this activity can be brought to life again when a second consciousness activates it and takes its schemata as cues for reciprocal meaning creation, for the gap filling, consistency building, and illusion making that we undertake when we read. The act of reading responds to the inert marks on the page as traces of activity that can be reanimated, and the neurobiological correlates of this miracle are the mirroring functions that the experiments linking language and motor action have identified.

The capacity of narrative to organize action no doubt similarly brings into play the relations between language and our motor system. This is, of course, a broad and speculative statement that would require much further experimental and theoretical work to test and refine, but the frequently attested notion (dating back at least to Aristotle) that narrative mimetically structures human action makes it a claim worth pursuing. In a recent reformulation of Aristotle's theory, Paul Ricoeur argues that the existential function of a "plot" is to organize events into structures of action that give narrative coherence to the flux and flow of temporal experience. The paradox of "emplotment," according to Ricoeur, is that it is "a synthesis of the heterogeneous"—creating "concord" from "discord," as I explained in the analysis in chapter 4 of the temporality of meaning creation—a synthesis that both is and is not faithful to the original experience that it restructures and reinterprets. The notion of mimesis as copying does not do justice, Ricoeur contends, to this "refiguration" of experience.[60]

Recognizing the resonances between linguistic and bodily action makes possible a different, more adequate model of mimesis that helps to explain how such restructuring can happen. To see language as a kind of action that sets in motion our motor capacities suggests an embodied, experientially based view of imitation, not as abstract, representational correspondence but as a lived resonance between original experience and its linguistic reactivation. These resonances play off each other in a back-and-forth manner, constituting a circle that, according to Ricoeur, makes narrative paradoxical: "temporality is brought to language to the extent that language configures and refigures temporal experience" (54). The temporality of language and the temporality of original experience can play off each other in this way because they are different modes of action. To view imitation as an interplay between kinds of action is to see mimesis not as inert epistemological copying but as embodied resonance, and two systems resonating interactively can have reciprocal effects on each other in a way that a unidirectional correspondence of a linguistic representation to the thing represented cannot.

Because of its links to our motor capacities, language can be viewed as a kind of "symbolic action," in Kenneth Burke's well-known formulation.[61] As such, language is a kind of action that can stage relations among other actions. Reading, in turn, is an activity that responds to the implied action in language, and this reactivation is the enabling condition of mimesis. Because the action embedded in language resonates with our motor capacities, it can organize actions into a structure that sets in motion in the reader a simulating response as if these actions were original. These are the sorts of resonances that a neurobiological investigation of narrative should explore.

This "as if" makes linguistic action "symbolic," in Burke's sense, and this "as" relation allows narrative to refigure experience in the way Ricoeur describes. Linguistic representation is a doubling, an action that may resonate immediately with the reader because it sets off correlative motor activity in the brain; but as a re-creation, a simulation, and a refiguration, it is only indirectly direct. This difference is the very condition of possibility of the mirroring mechanisms of the brain, a difference that is also fundamental to the distinction between story and discourse, between the events and their reconfiguration in a narrative design. This doubling is what makes narrative possible even as it prevents mimesis from ever being an exact copy. The play of the "as if" is more fluid, flexible, and variable than the point-to-point correspondence between a copy and what it represents, and that is why so many different kinds of narrative can lay legitimate claim to being realistic—why realism is a variable, historical set of conventions rather than a single representational ideal.

The "as if" of symbolic linguistic action also gives narrative the ability to feed back and provide structures for organizing our motor capacities. When Aristotle, in his famous definition of tragedy, describes it as "the imitation of an action," with a beginning, a middle, and an end, this perhaps seemingly obvious observation is nontrivial because it points to the capacity of symbolic linguistic action to shape and inform lived experience by its resonances with our motor capacities.[62] It may seem self-evident that a dramatic action must have a beginning, a middle, and an end, but this tripartite structure is not simply "there" in original unreflective experience. The "as if" of narrative refiguration of experience into this kind of structure gives action a particular shape that both is and is not what it originally was. This difference can be both instructive, showing us patterns that make sense of experience, and formative, constructing our worlds in particular shapes that may come to define the cultures we belong to. If dramatized action can refigure the structure of experienced action in these ways, the explanation lies at least in part in the resonances between the symbolic action of language and our motor mirror neurons, which make possible reciprocal relations between linguistic representations and lived experience.

Action in reading both is and is not the original action it reactivates and simulates, and this is one of the ways in which reading entails doubling. Poulet oversimplifies this relation when he defines reading as "identification," but his mistake deserves careful scrutiny because it illuminates how the brain mirrors action in reading and in life. According to Poulet, "When I read as I ought, i.e. without mental reservation, without any desire to preserve my independence of judgment, and with the total commitment required of any reader, my comprehension becomes intuitive and any feeling proposed to me is immediately assumed by me" (1215). Although he acknowledges that "there is no critical identification which is not prepared, realized, and incarnated through the agency of language" and "the mediation of words" (1217), he suggests that the most profound aesthetic experiences occur when this screen falls away:

> When reading a literary work, there is a moment when it seems to me that the subject *present* in this work disengages itself from all that surrounds it, and stands alone. Had I not once the intuition of this, when visiting the Scuola de San Rocco in Venice, one of the highest summits of art, where there are assembled so many paintings of the same painter, Tintoretto? When looking at all these masterpieces brought there together and revealing so manifestly their unity of inspiration, I had suddenly the impression of having reached the common essence present in all the works of a great master, an essence which I was not able to perceive, except when emptying my mind of all the particular images created by the artist. I became aware of a subjective power at work in all these pictures, and yet never so clearly understood by my mind as when I had forgotten all their particular figurations. (1221, emphasis in original)

This is an almost mystical out-of-body experience in which the difference between self and other is momentarily transcended. The opposition between ego and alter ego that paradoxically stands in the way of intersubjectivity even as it makes their relation possible has been miraculously overcome.

Such an experience of "I-thou" union shows our mirror neurons working almost too well. As Cristina Becchio and Cesare Bertone observe, the problem for the mirroring mechanisms of the brain "is not 'how is it possible to share the intentions of others,' but rather 'how can one distinguish one's own action/intention from those of other people.'"[63] If our mirror neurons fire both when we observe another acting and when we perform the same action ourselves, our brains would seem doomed to perpetual identity confusion. As Ramachandran notes, "out-of-body experiences" are neurobiologically interesting as evidence of "disruption to the inhibitory circuits that ordinarily keep mirror-neuron activity in check" (272). Sometimes these inhibitory mechanisms do indeed fail, as in cases of *echopraxia*, in which, as Gallese explains, patients "show an impulsive tendency to imitate other people's movements."[64] Nonpathological instances of imitative behavior are common in everyday life, as when others insist on completing one's sentences or when spouses and domestic partners take on each other's mannerisms.

Identity confusion is not more widespread in part because only a subset of cortical motor neurons show mirroring behavior (roughly 20% in the Parma monkeys), and they also spike less for observation than for action. As Iacoboni notes, "Mirror neurons embody both the interdependence of self and other—by firing for the actions of both—and the independence we simultaneously feel and require, by firing more powerfully for actions of the self" (133).[65] It is probably also the case, as Rizzolatti explains, that "braking mechanisms" in the frontal lobes inhibit the execution of acts we observe.[66] Such mechanisms fail to function in patients with "phantom limbs," who strangely report sensations in their missing body part when observing another person being touched on the corresponding place on his or her body. Ramachandran, who has conducted an oft-cited series of phantom-limb experiments, describes how this might work: "Perhaps the null signal ('I am not being touched') from skin and joint receptors in your own hand blocks the signals from your mirror neurons from reaching conscious awareness," inhibitory signals missing when the limb is absent. Without this braking mechanism, the patient's brain is confused and reacts as if the phantom part were actually being touched. He concludes: "It is the dynamic interplay of signals from frontal inhibitory circuits, mirror neurons . . . , and null signals from receptors that allow you to enjoy reciprocity with others while simultaneously preserving your individuality" (125).

These inhibitory signals have fallen away in Poulet's mystical experience of oneness with Tintoretto. Indeed, Poulet's moment of aesthetic transcendence makes it seem as if he has climbed out of his skin. This suggests that artistic form is in a perhaps metaphorical but also very real sense the "skin" of the aesthetic experience and that language is the "skin" of reading. Like skin, language is both a barrier and a connector between self and other, a medium with the capacity to carry signals between them that also demarcates their separation.[67] Even as language crosses the self-other divide by setting in motion resonances between the action embodied in derived intentionality and its reactivation in reading, it also marks the boundary between these modes of action. It is both the link and the divider between the author's originating meaning-making activity it embodies and our interpretive activity in constituting the text's intentionality in the particular gestalts we experience as we read. This duality embodies the doubleness of mediation that Poulet wishes to transcend but that is necessary for the reciprocal interaction of self and other that language makes possible.

Language enables this connection between self and other because it holds in readiness the patterns of activity that reading reanimates. But its various markers as only derived intentionality—not immediately available with the fullness of original experience but only schematized and incomplete in ways that give entry to the act of reading—are signals showing that the other we respond to as we read is paradoxically absent, not-there, even as we think this other's thoughts with an intimacy impossible in life. This doubling makes identification a more complex, paradoxical phenomenon than the oneness with another subjectivity that transports Poulet into Tintoretto's consciousness. The doubling of me and not-me in reading may seem miraculous and mysterious, but it is not a mystical out-of-body experience. It is, rather, an instance of the general, everyday experience of connection with and difference from others that the mirroring and inhibitory mechanisms of the brain enact across the various kinds of "skin" that join and separate us.

These mechanisms are evident in experiences of empathy, a kind of intersubjective relationship with which reading is sometimes compared.[68] In the words of Martin L. Hoffman, the preeminent psychologist in this field, "empathy is the spark of human concern for others, the glue that makes social life possible."[69] Although *empathy* is a notoriously slippery term, most psychologists would agree with Jennifer Pfeifer and Mirella Dapretto that the "experiential core of empathy" is "shared affect between self and other."[70] This sharing of affect is what Theodor Lipps tries to capture in his term *Ein-*

fühlung, sometimes translated as "identification," in which one "feels oneself into" the experience of another. Lipps's classic example is the anxiety a spectator may feel while watching a circus acrobat on a high wire, an experience of empathic identification in which the thrill of the performance is based on the spectator's vicarious sense of the tightrope walker's danger.[71] This example also shows, however, that the identification in empathy is not simply a merger of self and other but a paradoxical doubling of me and not-me. The cool, calm, and collected acrobat probably does not feel the same fear, anxiety, and excitement that spectators experience—otherwise he might become paralyzed and fall—even as the spectator's ability to enjoy the act depends on his or her not being in any real danger. In empathic identification one both does and does not feel what the other feels, and this doubling in turn is what makes possible the aesthetic re-creation of affective experiences (the circus performance is a case in point).

Such doubling is evident in various conundrums associated with the study of empathy. Psychologists point out, for example, that empathy is not the same as sympathy or compassion. Hein and Singer even note that "empathy can have a dark side," as "when it is used to find the weakest spot of a person to make her or him suffer, which is far from showing compassion with the other."[72] Some psychologists argue that such malevolence shows a lack of fellow feeling: "Psychopaths readily understand others' mental states, including their affective states, but lack the feeling of empathy."[73] But it is also sometimes observed that feeling another's emotional state can lead to "personal distress" rather than "empathic concern" and that this may widen rather than bridge the gap between self and other.[74] "Feeling as the other feels may actually inhibit other-oriented feelings if it leads us to become focused on our own emotional state," writes Daniel Baston. "To feel sorry for your friend you need not feel hurt and afraid too. It is enough to know that she is hurt and afraid."[75]

But even such knowledge is not by itself sufficient. For example, as Jean Decety notes, psychological "studies show that, in a competitive relationship, . . . pain in the competitor leads to positive emotions," whereas "observation of the other's joy results in distress."[76] This finding may be an instance of experimental psychology demonstrating what is intuitively obvious; still, the inverse feelings of competitors about one another's emotional states clearly show that empathic identification is not simple oneness with others but a doubling of me and not-me that cannot be predicted to have a morally straightforward result. This is one reason why, as Suzanne Keen rightly points out, claims for the humanizing effects of empathy in reading are overly simplistic.[77] Reading may make it possible for us to "live another life" from the inside, but there is no telling in advance what the consequences of such doubling will be.

These psychological paradoxes have a neurobiological basis, as various experiments on the neuroscience of pain suggest. "Although we commonly think of pain as a fundamentally private experience," Iacoboni observes, "our brain actually treats it as an experience shared with others" (124). In an oft-cited experiment, William Hutchison's laboratory found a set of neurons in the anterior cingulate cortex that fired not only in response to a painful pin-prick but also at the sight of someone else's finger being stuck by a needle.[78] In a corroborating experiment, Tania Singer and her colleagues did brain scans of participants who were given a painful electric shock and compared these with fMRI images taken when the same subjects were shown electrodes being attached to a beloved partner's hand and were told that this person would also be shocked. The scans showed identical activity in the insula and the ACC.[79] So I literally can feel your pain.

But this mirroring mechanism is also sensitive to individual differences that suggest the brain's simulation of the other's pain is a doubling of self and other and not a simple, one-to-one match. For example, in a paper titled "Love Hurts," Yawei Cheng reports that when participants adopted "the per-spective of a loved-one," brain scans "elicited greater activation in regions that belong to the pain matrix than adopting the perspective of a stranger."[80] This finding is consistent with other experimental results that show empathy is stronger when the other is perceived to be more like the self: "In children and adults as well as in monkeys, behavioral empathy is known to increase with greater similarity between oneself and the target, on the basis of such factors as species, personality, age, or gender."[81]

Past experience also makes a difference. In another brain-imaging experi-ment that Cheng conducted, acupuncturists who were shown pictures of needles being inserted into various body parts had less activity in the pain-related areas of the brain than did members of a control group. One interpre-tation of this result is that "empathic brain responses" decrease "if the empa-thizer is frequently exposed to pain-inflicting situations."[82] The belief of the acupuncturist in the healing powers of such therapy may also be a factor. In either case, empathy is not automatic but is subject to individual differences.

Men and women may empathize differently as well. Hein and Singer report that "one recent study showed that empathic brain responses in men but not in women were significantly weaker when the person in pain was judged as unfair, as compared with a person seen as fair and likable" (156). These differences at the neuronal level are the biological correlates of the doubling of ego and alter ego that makes us paradoxically transparent and opaque to one another. Mirroring mechanisms in the brain may allow me to feel your pain with perhaps surprising immediacy, but I don't necessarily feel it as you do. Even when I share your pain, I do so in my own way (and not yours).

These mechanisms are consistent with the phenomenological description of empathic identification in reading as a doubled staging of relations between the real me of the reader and the alien me who is called into being by the text and whose thoughts I think and feelings I feel as I read. The aesthetic experience of affects is a particular instance of the general phenomenon of simulating the emotions of others, which the mirroring mechanisms of the brain make possible. In reading, according to Iser, "'identification' is not an end in itself, but a strategem by which the author stimulates attitudes in the reader" (*Implied Reader* 291). Through various rhetorical devices, narrative strategies, and textual conventions, authors provoke simulations of thought processes, perceptions, and emotions in the reader that may be both familiar and strange. Such provocations and manipulations are only possible because reading is an act of doubling.

This doubleness is responsible for the paradoxical combination of immediacy and detachment that typically characterizes aesthetic emotions. The affect evoked in aesthetic experiences is structurally double—at times powerful, even overwhelming, as our own attitudes are taken over by different ways of feeling and perceiving, and yet always marked by a kind of absence, negativity, and distance because these artistically induced states are not our own original, immediate experience of the world but stagings of it, simulations, "as if" we were undergoing the experiences set in motion by the text. It is a characteristic of aesthetic emotions that they both are and are not the emotions they simulate, with a doubleness that is also found in the mirroring processes of empathy.

How visceral and corporeal are these simulations? Are they fully embodied or merely mental? These are interesting and important questions not only about empathy but also about aesthetic emotions. One oft-cited experiment that is relevant here has to do with disgust, a visceral and primitive feeling

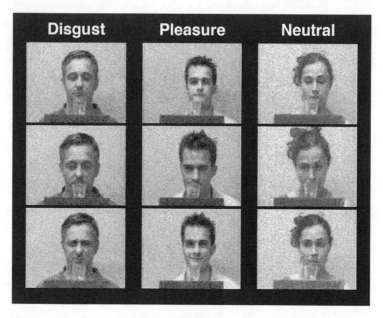

Figure 5.4. Reactions to disgusting, pleasant, and neutral smells. Still images from a video that fMRI subjects were shown of facial reactions to disgusting, pleasant, and neutral smells. Reproduced by permission from Bruno Wicker et al., "Both of Us Disgusted in *My* Insula: The Common Neural Basis of Seeing and Feeling Disgust," *Neuron* 40 (2003): 656 (fig. 1).

that probably goes back in evolutionary terms to the need to identify and avoid spoiled, rotten food. How did we learn to recognize from others' experience whether food was bad? Did we theorize their likely internal state from signs of disgust when they turned up their noses? Or is our empathic response more immediate and embodied? Bruno Wicker and his group hypothesized that "in order to understand the facial expression of disgust displayed by others, a feeling of disgust must occur also in the observer. This hypothesis predicts that brain areas responsible for experiencing this emotion will become active during the observation of that emotion in others."[83] In order to test this hypothesis, they performed brain scans on volunteers who were exposed to unpleasant smells and then compared these results with fMRI images taken when these same participants watched videos of subjects who sniffed pleasant, neutral, and disgusting substances (see fig. 5.4). The results showed activation at the same cortical site (the anterior insula) during the feeling of disgust stimulated by an unpleasant smell and during the observation of someone else's disgusted facial expression. Wicker concludes: "For disgust, there is a common substrate for feeling an emotion and perceiving the same emotion in others"; that is, "observing an emotion activates the neural representation of that emotion" (655).

This finding has reminded many mirror-neuron advocates of Antonio Damasio's as-if body loop. Recall Damasio's explanation of how "the brain can simulate certain emotional body states internally . . . when certain brain regions . . . directly signal the body-sensing brain regions" as if the body were signaling that it was undergoing some change of state: "What one feels then is based on that 'false' construction, not on the 'real' body state."[84] Although receptors in one's own nose are not sending signals of an unpleasant olfactory sensation, the same brain regions activate as if this were happening when they receive signals from the visual cortex that someone else's face is showing signs of disgust. As Damasio explains, the brain can simulate body states and "bypass" the body.[85] The paradox of this "as if" simulation is that the brain's re-creation of the body state in question both is and is not what it would be if we were undergoing the experience. The brain is sending signals to itself and not receiving signals from the body, but because the interacting brain regions are the same as if the body were in the loop, we feel the simulated emotion as if it were actually embodied.

Sometimes, however, the liminal location of the brain areas involved may blur the brain-body barrier even further. The involvement of the insula in both the experience of disgust and its observation can make our vicarious feeling of repulsion especially ambiguous, as Rizzolatti points out, because the insula is "the visceromotor integration center, which, when activated, provokes the transformation of sensory input into visceral reactions" (Rizzolatti and Sinigaglia 189). We may feel nauseous when we see another's disgusted facial expression, but we don't vomit (as we might if the disgust were fully and immediately embodied). But because of its mediating role connecting brain and body, the insula can send signals to the body that set off a visceral reaction and cause physical discomfort as if a nauseating stimulus were actually present. An "as if" simulation can have real, corporeal effects when the brain tells the body to react even though there is no stimulus there.

Similar analyses can be made of fear, which is associated with the amygdala (another important brain-body mediator), as well as other emotional sensations. All emotions cross the brain-body divide, and all can consequently be simulated in as-if loops, which may trick the brain into thinking signals are coming from the body, and they may do this so well that the brain signals the body to respond. Unlike disgust and fear, which are strongly linked to particular cortical sites, emotions typically involve not one exclusive area but what Bastiaansen calls "a mosaic of brain regions" that interact in complex networks.[86] No matter how they are organized in the cortex, however, the networks for emotions typically entail an interaction between what Damasio calls the "upstairs" regions of the neocortex and the "basement"—the "high and new" parts of the brain, which are associated with rationality and complicated mental functions, and the "low and old" areas, devoted to homeostasis and basic functions of bodily regulation that we share with more primitive species (see *Descartes' Error* 128). The as-if body loop fools the brain into thinking that such an interaction is occurring, but the brain may then send signals through the thalamus and the brain stem to the body that induce sensations related to the simulated corporeal state.

The ambiguities of the as-if body loop are evident in some of the characteristic contradictions and complications of aesthetic emotions. Applying Damasio's framework, Freedberg and Gallese argue that "when one observes pictures that arouse strong responses such as fear, the body is bypassed (for the most part, we do not actually run away, although we might), and the brain—within 'simulation mode'—reproduces the somatic states seen in or implied by the painting or sculpture, 'as if' the body were present" (201). "For the most part" is a crucial qualification. We probably won't run away from a frightening picture in a museum, but we might jump when we see a snake bite someone in a movie (I do, but my wife doesn't—although she has enough empathy, as well as sympathy, to know what's coming and hold me down in my seat). We may cry at the opera (as I have also been known to do) when the soprano sings a particularly heartbreaking aria, and some people have even been seen to break into tears in front of paintings (although less so now, James Elkins reports, than in the past).[87]

If different arts have different capacities to elicit emotional responses, that must be at least in part because of the way their sensory modalities set off resonances in the mirror regions in the brain. These modalities in turn may have different potentialities for triggering visceral, corporeal responses through as-if body loops. Such variations are what Nietzsche famously endeavors to explain in his analysis of the cool, detached "Apollonian" qualities of sculpture as opposed to the "Dionysian" abandon and self-loss that music can inspire, an opposition mediated, in his view, by tragedy.[88] It is certainly the case that the same scene represented in a painting or in a movie may have different effects and that hearing a love song performed can stimulate different emotional reactions than reading a poem silently. Relatively indirect in the sensory-motor responses it calls upon (if one compares the symbolic action embedded in language with the vivid, visually immediate action of a movie or the embodied resonance of a singing voice), reading offers a perhaps more mediated and less visceral experience than other arts (but I still weep with sympathetic grief every time I read Ralph Touchett's deathbed scene in *Portrait of a Lady*).[89]

The mechanisms underlying these differences are no doubt complicated, with many contributing factors, but such differences are possible because the brain can simulate the body with varying degrees of immediacy. This variable immediacy has to do with the ways in which empathic, emotion-sharing experiences may involve either a simulation confined to the cortex alone or an "upstairs-downstairs" interaction across the brain-body divide. Some of these differences also depend on differences in the participants, consistent with what the empathy experiments show about how experience and attitude may modulate identification. That is why sentimental reactions can vary culturally and historically (so that it may very well be that we no longer weep so much at pictures as we once did) and why some people jump at films with snakes (and others don't). There is no single aesthetic emotion, but all emotions induced by the various arts are "aesthetic" in the sense that they set in motion "as if" simulations that may be variously embodied. These differences in how emotions can be staged aesthetically exploit the variations of the "as if" that make empathically re-created bodily emotions ambiguously both identical to and different from the original. Just as I do and do not share your pain, so the pain I feel when I read about another's tragedy both is and is not real pain.

Similar mechanisms of doubling are at work in imitation in life and art. The discovery of mirror neurons prompted renewed interest in experiments on infant imitation on the assumption that newborns' capacity to copy observed behavior must be based on mirroring mechanisms hard wired in their brains. Although probes cannot be attached to newborns' neurons, observations of their ability to imitate adults within hours and even minutes of birth provide compelling evidence of the brain's innate mimetic capacities. As Andrew N. Meltzoff, the pioneering investigator of infant imitation, points out, "There are no mirrors in the womb," and "infants can see the adult's face but cannot see their own faces," and so the question is how they manage to "connect the felt but unseen movements of the self with the seen but unfelt movements of the other."[90] Infants can accomplish this act of translation almost as soon as they are born. Studying a sample of newborns as young as forty-two minutes and no older than seventy-two hours, Meltzoff found that "the newborns' first response to seeing a facial gesture is activation of the corresponding body part"—"an activation of the tongue," for example, "when they see tongue protrusion" (492). Infants' mimetic abilities steadily expand in the early days and weeks of life: "12–21-day old infants could imitate four different adult gestures: lip protrusion, mouth opening, tongue protrusion, and finger movement," and "infants confused neither actions nor body parts" (Meltzoff and Decety 492; see fig. 5.5). Meltzoff concludes: "Imitation is innate in human beings, which allows them to share behavioral states with other 'like me' agents" (492).

How can infants accomplish these acts of translation? Meltzoff observes that "human acts are especially relevant to infants because they look like the infant feels himself to be" (497). This feeling of likeness is in all probability based on neuronal resonances between internal states and signals received from outside. Gallese proposes that the firing of infants' mirror neurons in response to observed behavior sets in motion a process of "cross-modal mapping" of "visual information . . . into motor commands for reproducing it."[91] The infant makes sense of what it sees by undertaking motor acts that, because of the firing of its mirror neurons, are internally felt to be equivalent to what it observes.

This is an act of translation, and not a simple instance of one-to-one copying, because it is a setting equivalent of what both is and is not the same (another neuronal example of the hermeneutic "Gleichsetzen des Nicht-Gleichen"). This capacity for mapping is characteristic of many brain functions, across a variety of sensory modalities. Vision, hearing, smell, and touch all work by mapping inputs from various sources onto different brain regions. A process basic to how the brain establishes relations between signals and creates consistent patterns, area-to-area mapping also allows coordination of signals from different sensory inputs (as in the interaction of the aural and visual pathways in reading). In infant imitation, the resonances of mirror neurons probably trigger a similar cross-modal mapping between external observation and internally sensed capacities for action. That is, the infant correlates visual signals with analogous patterns of motor sensation and

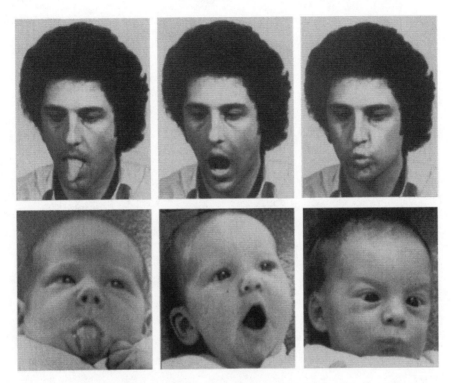

Figure 5.5. Images of infant imitation. Photographs of infants from twelve to twenty-one days old imitating the facial expressions of an adult. Reproduced by permission from Andrew N. Meltzoff and Jean Decety, "What Imitation Tells Us About Social Cognition: A Rapprochement Between Developmental Psychology and Cognitive Neuroscience," *Philosophical Transactions of the Royal Society London B* 358 (2003): 492 (fig. 1).

translates what it knows from experiences of the movement of its own body, even within the womb, to make sense of similar but novel, unprecedented information—actions by an observed protruding tongue, for example, which resonate with its own motor neurons in the area of the cortex that controls movements of this part of the body. A similar correlation occurs whenever we coordinate signals from sight and touch. This is perhaps the earliest instantiation of the hermeneutic circle, whereby we understand the unfamiliar by extending the familiar. The "as" structure of interpretation—how we make sense of something unfamiliar by analogizing it with what we already know—is at work from the outset of life in the brain's capacity to make these intermodal connections.

Such translation is crucial for the development of social cognition, Gallese points out, because of "the computational difficulties inherent in any type of interpersonal mapping, due to the different perspectives of demonstrator and imitator": "I always need to translate [the] external perspective of the demonstrator into my own personal body perspective" ("Manifold Nature of Interpersonal Relations" 519). Infant imitation is evidence of a primary intersubjectivity that precedes the development of the capacity to entertain beliefs about other minds, but the brain's mapping capacities at work in these translations are also active in the simulation of others' states and in theorizing about their points of view. There is continuity in the development of our capacities for intersubjective understanding because, as Gallese explains, "our cognitive system is capable of conceiving an 'abstract' multimodal way to map apparently unrelated sensory sources of information well before the development and mastery of language (the cognitive tool of abstraction *par excellence*) and of more sophisticated forms of social interaction" (518). Infants have this capacity for transmodal mapping right from birth because of the doubling that the resonances of mirror neurons make possible. Infant imitation entails doubling because the other's signals to which the infant responds both are and are not equivalent to what he or she has already experienced. The capacity of intermodal mapping to establish equivalences that are not simple repetitions is basic to the organism's ability to learn, develop, and expand its mastery of itself and its world.

The translational capacities of imitation depend crucially on the "as if" in ways that make possible play and fictional creation and help explain why humans can learn from make-believe.[92] Imitation can lead to improvement in an action or skill only if it is not merely one-to-one copying. For doubling to be effective and productive, the difference is as important as the similarity. Meltzoff reports that "young babies correct their imitative behavior," which is only possible with "an active comparison and lack of confusion between self and other" (Meltzoff and Decety 494). Further, eighteen-month-old infants can distinguish between successful and unsuccessful attempts to do something and "choose to imitate what we meant to do, rather than what we mistakenly did do" (496). Infant imitation is not simply an attempt to match observed behavior, then, but entails an active play with difference. The contrast, as well as the comparison, between what the infant sees and feels, between the original and the copy, is key to its ability to interact productively with the other in ways that allow for learning and improvement. When the adult misperforms, the infant knows to act "as" the other intended to (and not as she actually did). When the infant's acts fall short of what it aims to do, it can try again and attempt to act differently, and more "as" it wanted to all along. The infant acts "as" it sees the other acting, and the space between like and not-like in the "as" allows difference to play. Make-believe in play enacts these "as" structures, and play is crucial in infant learning because it allows experimentation with difference.

Reading continues the play of the "as" that begins with infant imitation. Aesthetic experiences that stage behaviors "as if" they were really happening are typically not only pleasurable diversions but also occasions for learning because of the "as" relations basic to imitation. The "as" of the "as if" has famously been a target of criticism, from Plato onward, because it is fundamentally deceptive and misleading (hence Sir Philip Sidney's classic defense that "the poet, he nothing affirms, and therefore never lieth").[93] But the "as" is productive, as infant imitation suggests, because it allows acts of simulation, and simulating what we are not allows us to explore and perhaps become what we might want to be. The play of negation (the "not" in the "as") is productive even in infant imitation because difference makes change possible. Staging "as" relations between me and not-me in reading continues this play of differences by juxtaposing opposing ways of thinking and feeling and carving out divisions within my personality that can give rise to thought or provoke changes in how I experience the world. Fictions can instruct as well as please not in spite of but because of the play of difference set in motion by the "as if" structure.

In art as in life, imitation is inherently social in ways that observation of infant development interestingly suggests. Infant imitation has the quality of play not only because of its "as if" dimension but also due to its to-and-fro dynamic. The to-and-fro of play is crucial both to imitation and to the doubling of self and other in intersubjective interactions. Meltzoff reports that "adults across cultures play reciprocal imitative games with their children" (Meltzoff and Decety 494). The back-and-forth of self-other reciprocity seems to be basic to the pleasure and learning of imitative experiences even in infancy, and this is especially evident when the tables turn and the imitated party does the imitating. According to Meltzoff, "Even very young children are attentive to being imitated": "They looked longer at the adult who was imitating them; smiled more at this adult; and most significantly, directed testing behavior at that adult, . . . performing sudden and unexpected movements to check if the adult was following what they did" (494).

This reciprocity of self and other enacts the paradox of the alter ego, the paradox of the other's independence and difference from me despite his or her intimate relatedness and responsiveness to me. As Meltzoff observes, "By 14 months, infants undoubtedly know that adults are not under their total control, and part of the joy of this exchange is the realisation that although the infant does not actually control the other, nonetheless the other is choosing to do just what I do. Together these two factors may help to explain why older infants will joyfully engage in mutual imitation games for 20 minutes or more—much longer and with greater glee than watching themselves in a mirror" (495). The reverse side of this playful exploration of self-other doubling is what Ramachandran describes as the "mental aloneness" of autism, where "an absence of emotional empathy for others" goes hand in hand with an inability to engage in to-and-fro exchanges: "autistic children express no outward sense of play, and they do not engage in the untrammeled make-believe with which normal children fill their waking hours."[94]

Not surprisingly, but in ways that have important implications for reading and aesthetic experience, language development depends on the to-and-fro play of infant imitation. Iacoboni reports that "the more a toddler plays imitation games, the more the same child will be a fluent speaker a year or two later" (50). The reasons for this connection are many and complex, but chief among them is that language is not only a logical and grammatical structure but also an embodied social practice that is exemplified by conversation. Denying that language "can be essentially reduced to formal constructs such as grammar," Iacoboni asserts: "The real question to ask is, How do people talk?" And he contends that "a salient feature of typical conversations that is ignored by traditional linguists is turn-taking."[95] As he explains, "Both the words and the actions in a conversation tend to be part of a coordinated, joint activity with a common goal, and this dance of dialogue is natural and easy for us" because it is "the kind of social interaction that mirror neurons facilitate through imitation" (*Mirroring People* 98).[96] The to-and-fro "dance" of conversation is one of the ways in which, according to Rizzolatti, "the mirror neuron system" helps to establish "a common space of action," a space of "reciprocal interaction" between my body and another's (Rizzolatti and Sinigaglia 154). The to-and-fro play of infant imitation improves the child's competence with the cognitive mapping necessary for negotiating such interactions.

Reading, especially literary reading, can be a playful to-and-fro because we experience similar doubling when we think and feel the thoughts and feelings of others. The derived intentionality embedded in written language is an invitation to engage in a "dance of dialogue." As Iser explains, there is not "a one-way incline from text to reader" but a "two-way relationship" in which the work's meaning depends on the reciprocal interactions of both parties, interactions that can occur in ways that transcend what an author may have originally intended or what the reader may have expected, hence the pleasure or frustration of surprise that is a typical aspect of aesthetic experience (*Act of Reading* 173).[97]

Jean-Paul Sartre describes the paradoxical to-and-fro of reading as an exchange of "demands" in which, reminiscent of adult-child imitation games, both parties take turns being in control: "Each one trusts the other; each one counts on the other, demands of the other as much as he demands of himself" in a kind of "dialectical going-and-coming; when I read, I make demands; if my demands are met, what I am then reading provokes me to demand more of the author, which means to demand of the author that he demand more of me. And, vice versa, the author's demand is that I carry my demands to the highest pitch" (49, 50). Sartre describes reading as "a pact of generosity" between author and reader because of this potentially mutually beneficial to-and-fro, which at its most intense brings out the most that both sides are capable of (and more than either could do on his or her own). The intuitive sense of "dialectical going-and-coming" that joins the reader in a "pact" with the absent author may seem mysterious and ineffable (Sartre describes the work of art as "a peculiar [spinning] top which exists only in movement" [34]), but the play of reciprocal appeals and demands in reading extends fundamental processes of mirroring and turn-taking that have their origins in infant imitation. The "dance" between author and reader is an embodied, materially grounded phenomenon that is based on the brain's capacities for intersubjective, intermodal mapping.

As a doubling of self and other, reading is a fundamentally collaborative process. It is, as such, a prime example of the "shared intentionality" that Michael Tomasello and other neurobiologically oriented cultural anthropologists identify as a unique human ability that gave rise to culture. What Tomasello calls "'we' intentionality" is the capacity for "participating in collaborative activities involving shared goals and socially coordinated action plans (joint intentions)."[98] The fundamental "skills of cultural cognition" made possible by shared intentionality begin with parent-infant "proto-conversations" that involve "turn-taking" and "exchange of emotions" (681), and they culminate in what is known as the "ratchet effect" of cumulative cultural evolution. Thanks to the human ability to engage in collaborative activity, culture can "ratchet up" specieswide change more rapidly and more broadly than biological evolution could accomplish. "Collective activities and practices, . . . often structured by shared symbolic artifacts," facilitate the "transmission across generations" of knowledge and skills (675). The socially induced neuronal recycling through which the brain learns to read is a primary example of the ratchet effect, and reading is a crucial vehicle of specieswide cross-generational transmission not only of the contents of culture but also of its forms. The centrality of reading and the lettered humanities to the transmission of culture is due to their facilitation of "'we' intentionality," not only through *what* is transmitted by means of writing and literature but also through *how* the actual practice of reading and writing develops our ability to collaborate with others. The to-and-fro play of reciprocal social interaction staged in aesthetic experience contributes powerfully to cultural evolution by enhancing shared intentionality.[99]

This is not to say, however, that the play of reading is inherently benevolent and socially productive, any more than our mirror neurons make us essentially moral beings. This is the fallacy of Ramachandran's description of mirror neurons as "Gandhi neurons" inasmuch as "they blur the boundary between self and others" (124). Although Patricia Churchland wants similarly to identify "the neural platform for moral behavior" (by which she means the neurobiological mechanisms through which the sphere of "caring" gets extended beyond "kin and kith"), she warns that "the platform is only the platform; it is not the whole story of human moral values" (3). It is also the case that imitation has two moral faces: it can enhance collaborative social interaction, but it can also provoke and promote violence. Iacoboni worries that "mirror neurons in our brain produce automatic imitative influences of which we are often unaware and that limit our autonomy by means of powerful social influences"; as an example he points out that "exposure to media violence has a strong effect on imitative violence" (*Mirroring People* 209, 206).

The effects of represented acts of violence are not necessarily noxious, however, for the very reason that their "as if" status as simulations gives the recipient room to maneuver. The mirroring responses of our neurons do not predetermine how we will react. Surveying evidence that "children who watch violent TV are at risk to become overly aggressive as adults," Richard Gerrig and Philip Zimbardo note that this depends on several factors that influence "observational learning," for example, whether the behavior is rewarded and reinforced, whether the model is viewed "positively, liked, and respected," whether the model is perceived as similar to the observer, and whether the behavior is within the observer's "range of competence."[100] Observational learning of aggression is not an automatic response, then, but is an "as" relation that may vary according to how the behavior is received, understood, and processed by the recipient. A doubling of me and not-me characterizes an observer's response to represented violence. Doubling is an inherent feature of all imitative behavior, and doubled relationships leave open variable possibilities of response. While endorsing research by Craig Anderson that suggests that "exposure to violent video games is a causal risk factor for increased aggressive behavior," Rowell Huesmann consequently acknowledges that "'increases the risk' . . . does not mean 'determines'": "for many exposed individuals, no detectable change in behavior will occur."[101] The analogy behind Huesmann's assertion is, however, biologically questionable: "The same statements can be made about most public health threats, including exposure to cigarette smoke and lead-based paint. The probability of lung cancer or intelligence deficits is increased by exposure but is not guaranteed" (179). The neurobiological mechanisms of imitation based on the resonances of mirror neurons are not biochemical reactions to environmental poisons, as with smoking and lead paint. The variability of imitative aggressive responses is not a question of the physiology of toxic exposure but has to do with the "as" relation of mirroring, and this increases its contingency and unpredictability.

This variability is reflected in dissents to the consensus about representational violence. Christopher Ferguson and John Kilburn question whether violent video games (VVGs) necessarily promote aggressive behavior: "As VVGs have become more popular in the United States and elsewhere, violent crime rates among youths and adults in the United States, Canada, United Kingdom, Japan and most other industrialized nations have plummeted to lows not seen since the 1960s."[102] They warn that "there are real risks that the exaggerated focus on VVGs, fueled by some scientists, distracts society from much more important causes of aggression, including poverty, peer influences, depression, family violence, and Gene X Environment interactions" (177). It should be noted, however, that some of these causes (e.g., aggressive behavior among peers and in the family) are also imitative effects. Similarly, in a recent review of the evidence, Daphne Bavelier cautions that the effects of violent video games "may indeed be fleeting rather than constituting true learned aggression effects" because she finds no correlation in the experimental evidence between "aggressive cognition" and "desensitization" and "being a regular player of violent video games."[103] This matter has been difficult to settle because imitative responses are contingent rather than automatic and are capable of inhibition or redirection.

The effects of imitation may not be preordained, but the evidence suggests that mirroring is not purely benevolent or benign, and it can inspire conflict and competition as much as it can support collaboration and mutually beneficial reciprocity. The anthropologically oriented literary theorist René Girard powerfully warns, for example, that what he calls "mimetic desire"—mirroring the desires of others and configuring our wants and wishes based on our envy of their perceived fulfillment—can spawn cycles of violence fueled by rivalry, jealousy, and resentment that only the sacrifice of a scapegoat can end.[104] Our biologically based abilities to imitate one another and engage in collaborative activities can cause horrors as well as wonders.

The secret of morality is not to be found in our neurons, and reading by itself will not make us better people. The paradox of the alter ego makes both conflict and care fundamental possibilities of human existence.[105] Given the evidence of both Machiavellian evil and selfless benevolence in human behavior, it would be surprising if the brain were intrinsically wired for either violence or compassion. Similarly, if aesthetic experience were always and necessarily morally uplifting and socially progressive, that would be hard to square with all too ample evidence that artists and humanists can behave badly and that cultural accomplishment can go hand in hand with brutal exploitation. What we need from both neuroscience and the lettered humanities is, not politically correct programs for social and moral improvement, but, rather, explanations from their different perspectives of the mixed picture of weal and woe that is repeatedly evident in human history. At least part of that story is to be found in the wonderfully and horribly disparate possibilities for how self and other can relate to each other that are inherent in the paradox of the alter ego and the capacity for doubling of the social brain.

Epilogue

I probably would not have written this book if I hadn't spent a dozen years as a dean.[1] That may sound odd, since doing administrative work is often (rightly) regarded as mind-dulling drudgery, and time spent in an administrative job is time not available for research and teaching. But I learned more about science during my decanal years than I ever did in high school or college. This was especially true of my experience as dean of arts and sciences at Stony Brook, a university known for its strengths in physics, mathematics, and the biological sciences (two of the departments that reported to me also had faculty from the medical school). A department chair, often a world-class scientist, would come into my office wanting money for a faculty position or for some research initiative, and persuading me of the merits of the case required giving this English professor enough of an education in the field that I could understand and evaluate the request. Recruiting and retaining faculty, building academic programs, and reading promotion and tenure cases in fields far from my own opened windows into areas I never would have known about if I had stayed in the professorial ranks. (Among other things, I learned how to read a scientific paper critically even if the technical details were beyond my competence.)

Rubbing shoulders every day with practicing scientists in ways that I had not done before (and have not done since) introduced me to ways of thinking and disciplinary practices that most humanists are unfamiliar with. Part of what is different between the humanities and the sciences is just that we know different things, and the fun of talking to scientists was learning about matters I was unaware of. But the sciences and the humanities are also different cultures, as C. P. Snow famously pointed out, and spending time outside of my disciplinary world exposed me to some assumptions, attitudes, and customs peculiar to this "other" community that have implications for the relation between neuroscience and aesthetics.

You know you are in another culture when familiar words don't function the same way. I still remember my surprise when I first heard a scientist use *reductionism* as a term of praise. English professors are accustomed to thinking that reductionism is an evil to be avoided at all costs. It is a tacit assumption shared by most humanists of whatever critical persuasion that no interpretation can fully do justice to the complexities of the poem or novel it attempts to explicate (an assumption elevated into doctrine by the New Critics as the "heresy of paraphrase"). One should at least demonstrate sufficient tact and grace to acknowledge this and, if possible, to make sure that the insight added by one's analysis makes up for the necessary oversimplifications of any commentary. Calling another interpreter's reading or method "reductionistic" is a serious charge. But for scientists, as I learned, "reductionism" is the name of the game. To scientists, no doubt, it will be surprising that this was a surprise to me, so natural are the conventions surrounding the use of this term in our different cultures. The goal of most science, after all, is to reduce complex phenomena to their simpler constituents, and part of the reason for the privileged status of physics and mathematics is that their methodological tools and explanatory concepts seem to promise to break the world down into its most fundamental elements and laws.

When I explained to colleagues in neurobiology, biochemistry, or ecology and evolution that the interactive, reciprocally formative processes characteristic of biological systems seemed to me to call for explanations reminiscent of the hermeneutic circle, according to which the whole is not reducible to the sum of its parts, they would quickly become nervous, fearing that I was introducing something mystical and nonmaterialistic, and would disavow the apparent similarities (politely, of course, because as their dean I still held the purse strings). It may be that biological systems are characterized by reciprocal, mutually determining interactions that aren't like linear, billiard-ball causality, but the nervousness of my scientist friends about my antireductionistic suggestions was a reminder that the goal was still to break the complex down to the simple and not to introduce anything extraneous that couldn't be accounted for in material terms.

Similar discipline-based anxieties and disagreements about reductionism are also often in evidence when the oft-discussed "hard problem" comes up. Can something like the aesthetic experience or the process of reading be reduced to the neural correlates of consciousness?[2] The answer you automatically give to that question will probably tell what culture you belong to (there are, of course, exceptions). I believe it is possible, however, to adopt the stance that John Searle calls *biological naturalism* and assert that "consciousness is a biological phenomenon caused by brain processes and realized in the structure of the brain" without committing the sins that humanists identify with reductionism.[3] As Searle argues, consciousness "is irreducible not because it is ineffable or mysterious, but because it has a first-person ontology and therefore cannot be reduced to phenomena with a third-person ontology" (567). Searle follows here Thomas Nagel's famous analysis showing that it is impossible for us to know what it is like to be a bat: "If the facts of experience—facts about what it is like *for* the experiencing organism [his bat or, in my question, the literary reader or anyone having an aesthetic experience]—are accessible only from one point of view [i.e., that of the organism having the experience], then it is a mystery how the true character of experiences could be revealed in the physical operation of that organism."[4] The ideology of science is committed to denying that anything is ultimately ineffable or mysterious, and its belief in reductionism follows. But this understandably makes humanists worry that what is distinctive about the experience will be lost. Such an explanation would miss precisely what it needs to account for: the "qualia," the technical term for what the experience is "like" in its full, immediate, first-person subjectivity.

But what if both are right, and also wrong? It is surely both good and bad to be a reductionist, depending on the explanatory work one is trying to do. What Francis Crick once called the "astonishing hypothesis," that is, "that to understand ourselves we must understand how nerve cells behave and how they interact," need not seem scandalous if the integrity of both of these opposing perspectives is recognized and if it is acknowledged, further, that neither can do the other's work and, one important step further, that each needs the other in order to do its work properly in some cases where their concerns overlap.[5] No one can read for you, and no one but you can have your aesthetic experience. That "my-own-ness" (*Jemeinigkeit*, in Heidegger's terminology) is what the humanist wants a nonreductionistic account to honor.[6] But reading and aesthetic experiences could not occur without brain processes (in the case of reading, for example, the neuronal recycling that repurposes visual invariant object-recognition cells so as to decipher graphic signs). The fact that these experiences and various underlying neuronal processes are correlated is something that can be documented and studied by the technical, experimental methods of neuroscience. But those methods are not the same modes of inquiry as one would want to use to understand and enhance the experience—to teach someone, for example, how to read a poem or a novel with greater insight into the literary conventions that are being invoked or violated, or to improve one's ability to recognize and understand melodic or harmonic patterns in a piece of music, or to interpret and respond to the patterns of spatial organization in a painting.

To show how two domains are correlated is not the same as demonstrating how one causes the other. Nor is it necessarily to privilege one domain as the only or best explanation of the other. In one sense, the brain activity explains what is going on in the aesthetic experience, but in another, equally valid sense, the aesthetic account of the lived experience of a poem, symphony, or painting explains what the brain activity means. Which explanation is primary is not intrinsic to either but depends on the work you want to do.

That is why there is what I have called an explanatory gap between the neuroscientific and aesthetic perspectives on phenomena like reading literary texts. Going back and forth across this "gap" gives neuroscientists and humanists useful things to talk about from their different perspectives. And they need to talk with each other (more than they typically do now) in order to do their work properly when their concerns cross. For example, neuroscientists should pay attention to the long history of accounts of aesthetic experience going back to Plato and Aristotle if they want to avoid serious methodological and theoretical errors (e.g., assuming that art is a univocal phenomenon that always invokes the same underlying reward mechanisms in the brain). Crosstalk with humanists would also be valuable if neuroscientists want promising suggestions about what to be looking for instead (how various aesthetic experiences of harmony and dissonance invoke local and global neuronal processes of particular kinds).

When humanists theorize, as we are wont to do, about whether a particular state of affairs is universal or historically and culturally relative, we should consult the relevant neuroscientific findings. These would show, for example, interesting evidence that linguistic signs are indeed arbitrary, contingent for their meaning on variable alphabetic and phonetic conventions, but that these differences are limited by certain apparently universal constraints that have to do with constant properties of the visual and auditory systems.[7] Similarly, often-heard claims about how radically reading processes have changed in recent centuries would need to be reformulated in light of the neuroscientific literature on the role of longstanding, evolutionarily stable visual processes in reading and interpretation.[8] Other instances of the relevance for literary theory of various experimental findings of neuroscience and, conversely, of literary theory for neuroscience have been pointed out in previous chapters. Because of the explanatory gap, neuroscientists and humanists have many things to learn from each other, even if neither will ever replace the other's work. That is not—or at least should not be—the goal.

The world would not be a richer place if the dreams of a "consilience" of the disciplines could ever be fulfilled and all knowledge could be unified by a few simple, basic laws. That is the ideal famously advocated, of course, by the biologist Edward O. Wilson, who embraces the aim of reductionism "to fold the laws and principles of each level of organization into those at more general, hence more fundamental levels": "Its strong form is total consilience, which holds that nature is organized by simple universal laws of physics to which all other laws and principles can eventually be reduced."[9] Wilson's assertion that this is a testable hypothesis that could be disproved scientifically is highly questionable, however, because it is not at all clear what would falsify it. That is one reason why the adherents and skeptics of reductionism often seem to be staring at each other across an unbridgeable divide, without any hope that a particular piece of evidence or knock-down argument could end their standoff.

More problematic, however, are the caveats and qualifications that Wilson introduces, which ultimately undermine the program they intend to shore up. For example, Wilson concedes that "at each level of organization, especially at the living cell and above, phenomena exist that require new laws and principles, which still cannot be predicted from those at more general levels" (60). So, as I have been arguing, there is indeed an explanatory integrity to different levels of organization and different approaches to knowledge that resists reduction to more basic laws and principles. That is why, as Wilson further acknowledges, "differences in validation criteria across the disciplines are accordingly vast" (63)—differences in what counts as a meaningful truth and how this can be tested according to the aims, assumptions, and methods of the particular explanatory approach. These are fatal concessions for the goal of "consilience," but they demonstrate the integrity and irreducibility of the epistemological work that is done on both sides of the explanatory gap.[10]

Similar implications follow from an example Searle offers to illustrate what he means by *biological naturalism*. To say that "conscious states are caused by neurobiological processes in the brain," he claims, "is analogous to saying that digestive processes are caused by chemical processes in the stomach and the rest of the digestive tract" (568). This may be true, but no one would suggest that a biological explanation of digestion could take the place of cooking lessons or the skill of a gourmet chef or the knowledge of a wine steward. Or to invoke another analogy I have previously employed, the ability of a professional baseball player to hit a fastball over the fence may be analyzable by the neuroscience of vision and the biology of motor coordination (which show it to be a nearly miraculous accomplishment, given the time scales of neuronal integration), but no one would advise the Red Sox designated hitter David "Big Papi" Ortiz to consult neuroscientists at Harvard or MIT when he is stuck in a slump. Hitting, pitching, and fielding can all be traced to neurobiological processes in the brain, but different kinds of knowledge and practical expertise than one finds in a scientific laboratory are required to improve performance on the field. The considerable neuroscientific research communities in Boston and New York may be competing in many areas, but nothing they can do will have any effect on the rivalry between the Red Sox and the Yankees.

As with these examples, so with reading and aesthetic experience: different domains of knowledge and practice require distinctive, domain-specific methods of understanding to do justice to their particular challenges, constraints, and possibilities. This suggests a model of interdisciplinarity other than what is typical in the humanities, and that too was one of my discoveries as a dean. When humanists do interdisciplinary work, we typically try to acquire as best we can knowledge about the insights and practices of the other discipline and then apply these to whatever interpretive project we are engaged in. My own academic degrees are all interdisciplinary on this model (history and literature as an undergraduate, modern thought and literature as a graduate student). Because interpreting texts is a lonely, individual experience of testing and revising hypotheses about the relation between parts and wholes, this model makes sense, and the other disciplines on which literary humanists have drawn—philosophy, anthropology, linguistics, political science—have often provided illuminating inspirations for productive hermeneutic activity.

Interdisciplinary inquiry in the sciences, however, typically involves cross-disciplinary collaboration that follows a different model. Although scientists sometimes find it profitable to learn methods and techniques from another discipline, they are usually fearful that amateurism and intellectual laxness might quickly endanger departures from one's area of disciplinary expertise. Thus, they are more likely to tackle a problem that goes beyond their own specialization by building teams of researchers from different disciplines with complementary intellectual strengths who together can solve problems none could address adequately in isolation. On this model, perhaps paradoxically, the best way to contribute to interdisciplinary inquiry is to have specialized command of a particular discipline's knowledge and procedures.

Although this book has been written following the humanistic model of interdisciplinarity (this humanist trying to learn enough neuroscience to speak competently about areas of mutual interest), its implications for neuroaesthetic work by scientists and humanists are consistent with the other model. Neuroscientists frequently go wrong, as I have argued, when they become amateur aestheticians. They need guidance about aesthetics and literary theory from humanists, who have wider and deeper knowledge about these issues. What humanists have most to offer neuroscience comes from our long engagement with core questions having to do with the creation and interpretation of works of art—the great variety of artistic forms that occur across different historical periods and cultural traditions, the range of aesthetic experiences to which these can give rise (from the harmonies promoted by unified, balanced structures to the disruptions provoked by dissonant, transgressive forms), and the widely differing, often conflicting theories that have been developed to account for this multiplicity.

Research in the humanities has gone back and forth over the decades between taking historical approaches to understanding this diversity and emphasizing the formal values and structures underlying it. If the danger of formalism is that it will wrongly universalize aesthetic phenomena that are historically contingent and variable, the risk for historicism is that it will forget aesthetic form in its quest for contexts to explain the social origins and political consequences of its many variations and changes.[11] After a long period in which cultural and historical approaches have dominated literary studies, a call to return to form has been heard in many quarters. Neuoraesthetics need not take sides in this dispute, however. Neuroscience needs both a historical appreciation of the heterogeneity and contingency of what can count as art and formal theories about the aesthetic experiences prompted by reading literary works.

History and form meet in the experience of reading. When literature plays with the brain, linguistic forms encountered in a novel, a poem, or a play by a particular, historically situated reader activate neuronal processes that are specific to the particular text and to the unique cortical wiring of the recipient but that also have transhistorical, cross-cultural, and evolutionarily longstanding properties that are related to fundamental features of neural anatomy and basic neurobiological processes. Instead of aligning itself with one or the other party in the ever-recurring debate between historicism and formalism, neuroaesthetics can instead demonstrate how their very conflict gives evidence of the duality of the brain as a universally occurring organ in our species that is remarkably open to adaptation, variation, and play. To do so, however, will require giving reading its due. The question how to read and what is involved in different ways of reading has long been central to the humanities, and it is a question with both formal and historical dimensions. Asking how the brain reads, and how literature plays with these processes, consequently calls on humanists to offer the sciences the benefit of our core knowledge and expertise. We too have much to gain from this exchange, not least of all a renewed appreciation of what it is we know that is distinctive to our work with literature.

Notes

CHAPTER 1: The Brain and Aesthetic Experience

1. V. S. Ramachandran, *The Tell-Tale Brain: A Neuroscientist's Quest for What Makes Us Human* (New York: Norton, 2011), 198.

2. See, however, John Hyman's critique of Ramachandran's notion of the "peak shift" effect in "Art and Neuroscience," in *Beyond Mimesis: Representation in Art and Science*, ed. Roman Frigg and Matthew C. Hunter (Heidelberg: Springer, 2010), 245–54.

3. On the complicated relation between the fixed and variable capacities of the brain, see Stephen Pinker's controversial, much-discussed book *The Blank Slate: The Modern Denial of Human Nature* (New York: Penguin, 2003). His avowedly opinionated chapter "The Arts" (400–420) is, however, perhaps the least valuable section of his book, and it shows the dangers of a scientist's presuming to pronounce unilaterally about the humanities ("They didn't ask me," he admits, "but by their own accounts they need all the help they can get" [401]). I happen to agree with many of his neuroscientific arguments about the limits of the brain's plasticity, but his sweeping generalizations about modern art and contemporary literary criticism, offered with much less precision, nuance, and rigorous argumentation than he feels it necessary to provide when discussing scientific issues, have understandably and unfortunately alienated many in the audience of humanists he needs to persuade.

4. In addition to my own books of practical criticism and the works of the Konstanz School theorists Wolfgang Iser and Hans Robert Jauss that I cite below, see the considerable scholarship produced by their *Nachwuchs* (next generation) in Germany and the United States, including Ulla Haselstein, Winfried Fluck, Gabriele Schwab, Anselm Haverkamp, Karlheinz Stierle, Rainer Warning, Evelyne Keitel, John Paul Riquelme, Brook Thomas, and Dale Bauer.

5. Norman N. Holland, "What is a Text? A Neurological View," *New Literary History* 33.1 (Winter 2002): 30. This essay argues, interestingly, that our sense of a text's autonomy, its existence "out there" in a "not-me" space, is accounted for by neurological processes in the brain whereby internal sensations and processes are projected externally because of the pragmatic evolutionary advantages of this illusion. From a less sympathetic perspective, Holland's skepticism about the usefulness of neuroscience to literary criticism is echoed by Raymond Tallis's ill-tempered diatribe, "The Neuroscience Delusion: Neuroaesthetics is Wrong About Our Experience of Literature and It is Wrong about Humanity," *Times Literary Supplement*, 9 April 2008. For interesting examples of practical criticism that take inspiration from neurobiology, see G. Gabrielle Starr, "Poetic Subjects

and Grecian Urns: Close Reading and the Tools of Cognitive Science," *Modern Philology* 105.1 (August 2007): 48–61; and Donald R. Wehrs, "Placing Human Constants within Literary History: Generic Revision and Affective Sociality in *The Winter's Tale* and *The Tempest*," *Poetics Today* 32.3 (Fall 2011): 521–91.

6. For an explanation of experimental brain-imaging technologies, see the chapter "The Tools: Imaging the Living Brain," in Bernard J. Baars and Nicole M. Gage, *Cognition, Brain, and Consciousness*, 2nd ed. (Amsterdam: Elsevier, 2010), 95–125. Briefly, in positron emission tomography (PET) a cyclotron measures metabolic brain activity by tracking the location of a radioactive tracer injected into the subject's bloodstream. A less expensive and more recent technology, functional magnetic resonance imaging (fMRI), exploits the magnetic properties of hemoglobin molecules to identify regions of the brain to which blood flow has increased because they have been activated. On the limitations of these technologies, see Alva Noë, *Out of Our Heads: Why You are Not Your Brain, and Other Lessons from the Biology of Consciousness* (New York: Hill & Wang, 2009), 19–24. In a cautionary analysis entitled "The New Phrenology?" Noë points out that "brain scans are not pictures of cognitive processes in the brain in action" but are at best indirect indications of activity, with "very low spatial and temporal resolution" (23–24).

7. See Uri Hasson et al., "Neurocinematics: The Neuroscience of Film," *Projections* 2.1 (2008): 1–26.

8. Colin McGinn, "Can We Solve the Mind-Body Problem?" *Mind* 98.391 (July 1989): 349, 354.

9. See Thomas Nagel, "What Is It Like to Be a Bat?," *Philosophical Review* 83 (1974): 435–50.

10. Adam Zeman, *A Portrait of the Brain* (New Haven, CT: Yale UP, 2008), 191.

11. Semir Zeki, *Splendors and Miseries of the Brain: Love, Creativity, and the Quest for Human Happiness* (Malden, MA: Wiley-Blackwell, 2009), 137.

12. For a similar argument, see John R. Searle, "Consciousness," *Annual Review of Neuroscience* 23 (2000): 557–78.

13. Martin Skov, "Neuroaesthetic Problems: A Framework for Neuroaesthetic Research," in *Neuroaesthetics*, ed. Martin Skov and Oshin Vartanian (Amityville, NY: Baywood, 2009), 11.

14. Hyman, "Art and Neuroscience," 261.

15. Evan Thompson, Antoine Lutz, and Diego Cosmelli, "Neurophenomenology: An Introduction for Neurophilosophers," in *Cognition and the Brain: The Philosophy and Neuroscience Movement*, ed. Andrew Brook and Kathleen Akins (Cambridge: Cambridge UP, 2005), 40.

16. Wolfgang Iser, *The Act of Reading: A Theory of Aesthetic Response* (Baltimore: Johns Hopkins UP, 1978), 166–67.

17. Kenneth Burke, "Terministic Screens," in *Language as Symbolic Action* (Berkeley: U of California P, 1966), 44–62.

18. Bruno Latour, *We Have Never Been Modern*, trans. Catherine Porter (Cambridge, MA: Harvard UP, 1993), 18. "When we are dealing with science and technology," he argues, "it is hard to imagine for long that we are dealing with a text that is writing itself, a discourse that is speaking all by itself, a play of signifiers without signifieds" (64). The inability of social constructionism to recognize the both-and of facts that are factual, not

illusory, even as they are historically contingent products of laboratory investigation is an example of the ethos of "purification" that, according to Latour, defines the "modern constitution" (see 10–15).

19. Examples of bad practice that plays fast and loose with scientific terminology unfortunately abound in what has come to be known as *affect theory*, an area that one might think would have potentially promising links to neurobiology. Phrases like these abound in *The Affective Turn: Theorizing the Social*, ed. Patricia Clough (Durham, NC: Duke UP, 2007); and *The Affect Theory Reader*, ed. Melissa Gregg and Gregory J. Seigworth (Durham, NC: Duke UP, 2010). An exception is Anna Gibbs, "After Affect: Sympathy, Synchrony, and Mimetic Communication," in Gregg and Seigworth, *Affect Theory Reader*, 186–205, although some of her speculations about "contagious behavior" could be more precise. By contrast, N. Katherine Hayles's work is exemplary, no doubt in part because her graduate training in chemistry and practical work as a chemist preceded her turn to literary study. See, e.g., Hayles, *How We Became Posthuman: Virtual Bodies in Cybernetics, Literature, and Informatics* (Chicago: U of Chicago P, 1999).

20. For a fuller analysis of these issues, see "Interpretive Conflict and Validity," the opening chapter of my *Conflicting Readings: Variety and Validity in Interpretation* (Chapel Hill: U of North Carolina P, 1990), 1–19, or the earlier published version, "The Conflict of Interpretations and the Limits of Pluralism," *PMLA* 98 (1983): 341–52.

21. For a more extensive analysis of these issues, see my essays "In Defense of Reading: Or, Why Reading Still Matters in a Contextualist Age," *New Literary History* 42.1 (Winter 2011): 87–113; and "Form and History: Reading as an Aesthetic Experience and Historical Act," *Modern Language Quarterly* 69 (June 2008): 195–219.

22. See Jonah Lehrer, "Coda," in *Proust Was a Neuroscientist* (Boston: Houghton Mifflin, 2008), 190–97. According to Lehrer's proposal, the "third culture" comprises scientists who write for a popular, nontechnical audience, and the "fourth culture" would be made up of both artists and scientists who seek "to discover relationships *between* the humanities and the sciences" (196, emphasis in original). For a thoughtful analysis of how the shared interests of neurobiology and the humanities in such matters as embodied cognition might lead to useful dialogue between the "two cultures," see Edward Slingerland, *What Science Offers the Humanities: Integrating Body and Culture* (Cambridge: Cambridge UP, 2008). Also see the chapter "Understanding and Truth in the Two Cultures" in my *Conflicting Readings*, 44–66.

23. As I explain shortly, the boom in interest in cognitive literary studies has focused on experimental psychology rather than hard-core neuroscience. An exception is G. Gabrielle Starr, who collaborates with a team of neuroscientists using fMRI technology to study art and the brain. See her interesting essay "Multisensory Imagery," in *Introduction to Cognitive Cultural Studies*, ed. Lisa Zunshine (Baltimore: Johns Hopkins UP, 2010), 275–91. As if to prove my point, hers is the only essay in that volume that makes connections to neuroscience. Outside of academic circles, Jonah Lehrer is a journalist who draws on his experience as an undergraduate working in a neuroscience lab to speculate about parallels between neuroscience and literature in his eminently readable book *Proust Was a Neuroscientist*. Two other important exceptions are Norman Holland's "neuropsychoanalysis" (as he terms his theory) and David S. Miall's "empirical" approach to reading, which is primarily psychological but is deeply informed by neuroscience. See Holland,

Literature and the Brain (Gainesville, FL: PsyArt Foundation, 2009); and Miall, *Literary Reading: Empirical and Theoretical Studies* (New York: Peter Lang, 2006). Also see the important collection *Neuroaesthetics*, ed. Skov and Vartanian (cited in n. 13 above). Skov is a Danish neuroscientist who first trained as a literary theorist.

24. On the cortical localization of color perception and facial recognition, see Zeki, *Splendors and Miseries of the Brain*, 65–72.

25. See R. Q. Quiroga et al., "Invariant Visual Representation by Single Neurons in the Human Brain," *Nature* 435 (23 June 2005): 1102–7.

26. David Keller, "Review of *Neuroaesthetics*," *British Journal of Aesthetics*, advance access, first published online 29 March 2010.

27. Jean-Pierre Changeux, *The Good, the True, and the Beautiful: A Neuronal Approach*, trans. Laurence Garey (New Haven, CT: Yale UP, 2012), 40, 11; Stanislas Dehaene, *Reading in the Brain: The Science and Evolution of a Human Invention* (New York: Viking, 2009), 310. The latter is Dehaene's summary of Changeux's argument in Changeux, *Raison et plaisir* (Paris: Odile Jacob, 1994).

28. V. S. Ramachandran and W. Hirstein, "The Science of Art: A Neurological Theory of Aesthetic Experience," *Journal of Consciousness Studies* 6.6–7 (1999): 16–17, emphasis in original. Ramachandran repeats and expands this argument in his most recent book, *The Tell-Tale Brain* (cited in n. 1 above), 192–244.

29. The *locus classicus* of this dispute is William Blake's angry refutation of Sir Joshua Reynolds's classical theory of beauty. See the selections from Reynolds's *Discourses on Art* (1768) and Blake's *Annotations to Reynolds' "Discourses"* (ca. 1808) in *Critical Theory Since Plato*, ed. Hazard Adams (New York: Harcourt Brace Jovanovich, 1971), 354–76, 402–12.

30. See Cleanth Brooks, "The Language of Paradox," in *The Well Wrought Urn* (New York: Harcourt, Brace & World, 1947), 3–21; and Paul de Man, "Semiology and Rhetoric," in *Allegories of Reading* (New Haven, CT: Yale UP, 1979), 3–19.

31. Martin Skov and Oshin Vartanian, "Introduction: What is Neuroaesthetics?" in Skov and Vartanian, *Neuroaesthetics*, 4.

32. Patrick Colm Hogan, "Literary Universals," in Zunshine, *Introduction to Cognitive Cultural Studies*, 37–60. Acknowledging some of the difficulty here, Hogan concedes that "universals . . . are not necessarily properties of all literary works" and "also need not apply to all traditions" but says they are features found "across (genetically and areally unrelated) languages with greater frequency than would be predicted by chance alone" (42). They are statistically significant generalities, then, and not "universals." Hogan also fails to consider the possibility that aesthetic criteria (like harmony or dissonance) might be radically opposed and mutually exclusive. On the problems of defining what is aesthetic, see the chapter "The Variability and Limits of Value," in my *Conflicting Readings*, 109–33.

33. Roman Jakobson, "Linguistics and Poetics," in *Style in Language*, ed. Thomas A. Sebeok (New York: Wiley, 1960), 350–77.

34. David S. Miall, "Neuroaesthetics of Literary Reading," in Skov and Vartanian, *Neuroaesthetics*, 237. A noteworthy exception to the lack of connection to contemporary neurobiological research, Miall has had a longstanding collaboration with a colleague in cognitive psychology at the University of Alberta and draws on neuroscience in his work. Apart from my skepticism about his proposal for demarcating literariness, this essay and his book *Literary Reading: Empirical and Theoretical Studies* (cited in n. 23 above) offer thoughtful, informed explorations of the intersections of literature and neuroscience.

35. Clive Bell, "The Aesthetic Hypothesis," in *Art* (1914; London: Chatto & Windus, 1949), 6, 8, 25.

36. Immanuel Kant, *Critique of Judgement* (1790), trans. James Creed Meredith, ed. Nicholas Walker (Oxford: Oxford UP, 2007), esp. 35–74.

37. See Alan Richardson and Francis F. Stein, "Literature and the Cognitive Revolution: An Introduction," *Poetics Today* 23.1 (Spring 2002): 1–8.

38. For examples of the best work of this kind, see Alan Richardson, *British Romanticism and the Science of Mind* (Cambridge: Cambridge UP, 2001); Nicholas Dames, *The Physiology of the Novel: Reading, Neural Science, and the Form of Victorian Fiction* (New York: Oxford UP, 2007); and Vanessa L. Ryan, *Thinking without Thinking in the Victorian Novel* (Baltimore: Johns Hopkins UP, 2012). Although he remains primarily a neurohistorian, Richardson explores connections between Romanticism and contemporary cognitive science in his most recent book, *The Neural Sublime: Cognitive Theories and Romantic Texts* (Baltimore: Johns Hopkins UP, 2010).

39. See, e.g., Lisa Zunshine, *Why We Read Fiction: Theory of Mind and the Novel* (Columbus: Ohio State UP, 2006); Zunshine, *Strange Concepts and the Stories They Make Possible* (Baltimore: Johns Hopkins UP, 2009); Alan Palmer, *Fictional Minds* (Lincoln: U of Nebraska P, 2004); and Patrick Colm Hogan, *Cognitive Science, Literature, and the Arts: A Guide for Humanists* (New York: Routledge, 2003).

40. Dehaene, *Reading in the Brain*, 257, emphasis in original.

41. Andrew Brook and Peter Mandik, introduction to Brook and Akins, *Cognition and the Brain* (cited in n. 15 above), 6–7.

42. Francisco J. Varela, "The Specious Present: A Neurophenomenology of Time Consciousness," in *Naturalizing Phenomenology: Issues in Contemporary Phenomenology and Cognitive Science*, ed. Jean Petitot et al. (Stanford, CA: Stanford UP, 1999), 267.

43. Thompson, Lutz, and Cosmelli, "Neurophenomenology" (cited in n. 15 above), 89, emphasis in original.

44. See Francisco J. Varela and Jonathan Shear, eds., *The View from Within: First-Person Approaches to the Study of Consciousness* (Bowling Green, OH: Imprint Academic, 1999).

45. See Varela, "Specious Present," 280–95; and Edmund Husserl, *The Phenomenology of Internal Time Consciousness* (1928), ed. Martin Heidegger, trans. James S. Churchill (Bloomington: Indiana UP, 1964). These connections also have important implications for the temporality of reading, as I explain in chapter 4.

46. For a concise survey, see my entry "Phenomenology" in *The Johns Hopkins Guide to Literary Theory and Criticism*, ed. Michael Groden and Martin Kreiswirth (Baltimore: Johns Hopkins UP, 1994), 562–66, as well as my article "Hermeneutics" in *Blackwell Encyclopedia of Literary and Cultural Theory*, gen. ed. Michael Ryan, vol. 1, *Literary Theory from 1900 to 1966*, ed. Gregory Castle (Malden, MA: Wiley-Blackwell, 2011), 236–46.

47. See Semir Zeki, "Brain Concepts and Ambiguity," in Zeki, *Splendours and Miseries of the Brain*, 59–98; and my *Conflicting Readings*. Chapter 3 below explores these parallels in detail.

48. For an analysis of why contemporary culturally and historically oriented literary criticism is suspicious of the experience of reading and why this disdain is mistaken, see my essay "In Defense of Reading" (cited in n. 21 above).

49. See Roman Ingarden, *The Literary Work of Art* (1931), trans. George G. Grabowicz (Evanston, IL: Northwestern UP, 1973); and Ingarden, *The Cognition of the Literary Work of*

Art (1937), trans. Ruth Ann Crowley and Kenneth R. Olson (Evanston, IL: Northwestern UP, 1973).

50. Wolfgang Iser, "The Reading Process: A Phenomenological Approach," in *The Implied Reader: Patterns of Communication in Prose Fiction from Bunyan to Beckett* (Baltimore: Johns Hopkins UP, 1974), 279–80. Also see Iser, *Act of Reading* (cited in n. 16 above).

51. Hans Robert Jauss, "Literary History as a Challenge to Literary Theory," in *Toward an Aesthetic of Reception*, trans. Timothy Bahti (Minneapolis: U of Minnesota P, 1982), 25–26.

52. These quotations are taken from Ramachandran, *Tell-Tale Brain*, 257; Antonio Damasio, *Self Comes to Mind: Constructing the Conscious Brain* (New York: Pantheon, 2010), 64, 87; and Daniel C. Dennett, *Consciousness Explained* (New York: Little Brown, 1991), 16.

53. Noë, *Out of Our Heads*, 47.

54. The *locus classicus* of this argument is Jacques Derrida's well-known analysis of Saussure's notion of the sign in "Structure, Sign, and Play in the Discourse of the Human Sciences," in *The Structuralist Controversy: The Languages of Criticism and the Sciences of Man*, ed. Richard Macksey and Eugenio Donato (Baltimore: Johns Hopkins P, 1970), 247–65. As Derrida famously notes, "Language bears within itself the necessity of its own critique" (254).

55. Jean-Pierre Changeux and Paul Ricoeur, *What Makes Us Think? A Neuroscientist and a Philosopher Argue about Ethics, Human Nature, and the Brain*, trans. M. B. DeBevoise (Princeton, NJ: Princeton UP, 2000), 14.

56. As Paul Ricoeur himself recognizes and explains in illuminating detail in his important book *The Rule of Metaphor*, trans. Robert Czerny (Toronto: U of Toronto P, 1977).

CHAPTER 2: How the Brain Learns to Read and the Play of Harmony and Dissonance

1. The classic account of this phenomenon is Stephen Pinker, *The Language Instinct: How the Mind Creates Language* (1994; New York: Harper, 2007). His Chomskyan claim that language is based on inborn, universal cognitive structures has recently been called into question, however, and is no longer the consensus view among neuroscientists of language. See Nicholas Evans and Stephen C. Levinson, "The Myth of Language Universals: Language Diversity and its Importance for Cognitive Science," *Behavioral and Brain Sciences* 32 (2009): 429–48, and the extensive accompanying "Open Peer Commentary," 448–92, esp. Michael Tomasello, "Universal Grammar is Dead," 470–71, but also the rebuttal by Stephen Pinker and Ray Jackendoff, "The Reality of a Universal Language Faculty," 465–66. Evans and Levinson argue that "language is a bio-cultural hybrid" and that "a property common to languages need not have its origins in a 'language faculty,' or innate specialization for language" (446, 439). My criticisms in subsequent chapters of Pinker's linguistic model align me with the critics of universal grammar. The neurological origins of language are no doubt more complex than the hypothesis of a "language instinct" can explain; nevertheless, for whatever reasons, language is a more "natural" acquisition than reading and is probably based on the long-term evolution of the human brain in ways that reading is not. Language is not as automatic as vision (although the visual cortex also will not develop unless it is used), but it is more inevitable than reading.

2. Steven Roger Fischer, *A History of Reading* (London: Reaktion Books, 2003), 16.

3. See Stanislas Dehaene, *Reading in the Brain: The Science and Evolution of a Human Invention* (New York: Viking, 2009), 144–47. My explanation of the neuroscience of reading is deeply indebted to this fascinating book. For a concise survey of the recent research and an analysis of its implications for the teaching of reading, see George G. Hruby and Usha Goswami, "Neuroscience and Reading: A Review for Reading Education Researchers," *Reading Research Quarterly* 46.2 (April–June 2011): 156–72.

4. For the debate about the VWFA, see Cathy J. Price and Joseph T. Devlin, "The Myth of the Visual Word Form Area," *NeuroImage* 19 (2003): 473–81; Laurent Cohen and Stanislas Dehaene, "Specialization Within the Ventral Stream: The Case for the Visual Word Form Area," *NeuroImage* 22 (2004): 466–76; and Price and Devlin, "The Pro and Cons of Labelling a Left Occipitotemporal Region: 'The Visual Word Form Area,'" *NeuroImage* 22 (2004): 477–79.

5. See, e.g., L. H. Tan et al., "Brain Activation in the Processing of Chinese Characters and Words: A Functional MRI Study," *Human Brain Mapping* 10.1 (2000): 16–27; and K. Nakamura et al., "Participation of the Left Posterior Inferior Temporal Cortex in Writing and Mental Recall of Kanji Orthography: A Functional MRI Study," *Brain* 123.5 (2000): 954–67. Also see Dehaene's summary of these experiments in *Reading in the Brain*, 97–100.

6. See T. A. Polk and M. J. Farah, "Functional MRI Evidence for an Abstract, Not Perceptual, Word-Form Area," *Journal of Experimental Psychology* 131.1 (2002): 65–72. Polk and Farah's brain-imaging experiment showed practically indistinguishable amounts of neuronal activity in the VWFA when reading words written in all uppercase or all lowercase letters or in a mix of the two, and they also found that oddly written words like *HoTeL* or *ElEpHaNt* triggered the same activity as normally written words.

7. Mark Changizi et al., "The Structures of Letters and Symbols Throughout Human History are Selected to Match Those Found in Objects in Natural Scenes," *American Naturalist* 167.5 (May 2006): E117. For a further discussion of Changizi's work, see Dehaene, *Reading in the Brain*, 176–79.

8. For a discussion of this experiment, see V. S. Ramachandran, *The Tell-Tale Brain: A Neuroscientist's Quest for What Makes Us Human* (New York: Norton, 2011), 108–9.

9. See the well-known Socratic dialogue on Cratylus's doctrine that names and things must be linked by a natural bond: Plato, *Cratylus*, trans. C. D. C. Reeve (Indianapolis: Hackett, 1998). For a recent analysis, see Francesco Ademollo, *The Cratylus of Plato: A Commentary* (Cambridge: Cambridge UP, 2011).

10. See Ferdinand de Saussure's classic analysis of the sign in his *Course in General Linguistics*, ed. Charles Bally, Albert Sechehaye, and Albert Riedlinger, trans. Wade Baskin (New York: McGraw Hill, 1966).

11. See, e.g., N. van Atteveldt et al., "Integration of Letters and Speech Sounds in the Human Brain," *Neuron* 43 (2004): 271–82. Also see Deheane's summary of the relevant research in *Reading in the Brain*, 104–9.

12. Price and Devlin, "Pro and Cons," 478.

13. Hruby and Goswami, "Neuroscience and Reading," 161, 157.

14. P. H. K. Seymour, M. Aro, and J. M. Erskine, "Foundation Literacy Acquisition in European Orthographies," *British Journal of Psychology* 94.2 (2003): 143–74. For a discussion of this study, see Dehaene, *Reading in the Brain*, 230–32.

15. Brian Boyd, *On the Origin of Stories: Evolution, Cognition, and Fiction* (Cambridge, MA: Harvard UP, 2009), 36. For an even-handed analysis of the sometimes insightful but sometimes highly questionable ways in which Darwin's thinking has been applied to interpret literary and cultural states of affairs, see George Levine, "Reflections on Darwin and Darwinizing," *Victorian Studies* 52.2 (Winter 2009): 223–45. A more skeptical critique is offered by Jonathan Kramnick, "Against Literary Darwinism," *Critical Inquiry* 37.2 (Winter 2011): 315–47. Also see Boyd's reply, "For Evocriticism: Minds Shaped to Be Reshaped," *Critical Inquiry* 38.2 (Winter 2012): 394–404.

16. See Claude Lévi-Strauss, *The Savage Mind* (1962), trans. anon. (Chicago: U of Chicago P, 1966), 16–33.

17. Stephen Jay Gould, *The Panda's Thumb* (New York: Norton, 1980), 13.

18. Semir Zeki, *Splendors and Miseries of the Brain: Love, Creativity, and the Quest for Human Happiness* (Malden, MA: Wiley-Blackwell, 2009), 33, 133.

19. Peculiar and illuminating impairments of this kind are, of course, the favored topic of the clinical neurologist Oliver Sacks in his popular books, such as *The Man Who Mistook His Wife for a Hat* (New York: Simon & Schuster, 1985).

20. See, e.g., M. H. Bornstein et al., "Perceptual Similarity of Mirror Images in Infancy," *Cognition* 6.2 (1978): 89–116.

21. See James M. Cornell, "Spontaneous Mirror-Writing in Children," *Canadian Journal of Experimental Psychology* 39 (1985): 174–79. Also see Dehaene's useful summary of the research on mirror symmetry in *Reading in the Brain*, 263–99.

22. See Laurent Cohen et al., "Learning to Read Without a Left Occipital Lobe: Right-Hemispheric Shift of Visual Word Form Area," *Annals of Neurology* 56.6 (2004): 890–94.

23. Andreas Bartels and Semir Zeki, "The Neural Correlates of Maternal and Romantic Love," *NeuroImage* 21 (2004): 1155–66.

24. Oliver Sacks, "Face-Blind: Why Are Some of Us Terrible at Recognizing Faces?" *New Yorker*, 30 August 2010, 41.

25. Anjan Chatterjee, "Neuroaesthetics: A Coming of Age Story," *Journal of Cognitive Neuroscience* 23.1 (2010): 56.

26. Judith H. Langois and Lori A. Roggman, "Attractive Faces are Only Average," *Psychological Science* 1.2 (1990): 115.

27. Hebb's Law is named for the neuroscientist Donald O. Hebb, who proposed it in his landmark book *The Organization of Behavior: A Neuropsychological Theory* (1949; Mahwah, NJ: Erlbaum, 2002). See Mark Bear, Barry W. Connors, and Michael A. Paradiso, *Neuroscience: Exploring the Brain*, 3rd ed. (Baltimore: Lippincott Williams & Wilkins, 2007), 733.

28. Thomas F. Münte et al., "The Musician's Brain as a Model of Neuroplasticity," *Nature Reviews/Neuroscience* 3 (June 2002): 475.

29. E. A. Maguire et al., "Navigation-related Structural Change in the Hippocampi of Taxi Drivers," *Proceedings of the National Academy of Sciences* 97 (2000): 4398–4403. On the functions of the hippocampus, see Mikko P. Laakso et al., "Psychopathy and the Posterior Hippocampus," *Behavioural Brain Research* 118.2 (29 January 2001): 187–93.

30. See D. W. Green et al., "Exploring Cross-Linguistic Vocabulary Effects on Brain Structures Using Voxel-Based Morphometry," *Bilingualism: Language and Cognition* 10 (2007): 189–99.

31. See Elkhonon Goldberg, *The New Executive Brain: Frontal Lobes in a Complex World*

(New York: Oxford UP, 2009), 238–39, where Goldberg also discusses the experiments involving taxi drivers and bilingualism.

32. See Bear, Connors, and Paradiso, *Neuroscience*, 693.

33. Valerie Gray Hardcastle and C. Matthew Stewart, "Localization in the Brain and Other Illusions," in *Cognition and the Brain: The Philosophy and Neuroscience Movement*, ed. Andrew Brook and Kathleen Akins (Cambridge: Cambridge UP, 2005), 28, 36, emphasis in original.

34. Rebecca Saxe, "The Unhappiness of the Fish: Understanding Other Minds That are Unlike Your Own," oral presentation, Harvard Cognitive Theory and the Arts Seminar, 22 March 2012.

35. See Ellen Spolsky's explanation of why the brain is messier and more chaotic than assumed, for example, by Jerry A. Fodor's influential model of a "central controller" that organizes a "modular mind" that knows the world in different, distinguishable, but interdependent ways in her book *Gaps in Nature: Literature, Interpretation, and the Modular Mind* (Albany: SUNY P, 1993), esp. 34, 38. This is one reason for the decline in the dominance of computer models to explain how the brain works. Also see Hubert L. Dreyfus, *What Computers Still Can't Do: A Critique of Artificial Reason* (Cambridge, MA: MIT P, 1992). Given the evidence for the localization of brain function, however, Spolsky wants to retain the notion of "modules," while abandoning the "controller."

36. Francisco J. Varela, "The Specious Present: A Neurophenomenology of Time Consciousness," in *Naturalizing Phenomenology: Issues in Contemporary Phenomenology and Cognitive Science*, ed. Jean Petitot et al. (Stanford, CA: Stanford UP, 1999), 274, 272.

37. Ramachandran, *Tell-Tale Brain*, 55.

38. See Dehaene's discussion of these and other amusing examples of semantic ambiguity in *Reading in the Brain*, 109–13. In the next chapter I discuss more extensively the implications of not only linguistic but also visual ambiguity for the neuroscience of understanding.

39. Marcos Nadal et al., "Constraining Hypotheses on the Evolution of Art and Aesthetic Appreciation," in *Neuroaesthetics*, ed. Martin Skov and Oshin Vartanian (Amityville, NY: Baywood, 2009), 123.

40. Dahlia W. Zaidel, "Brain and Art: Neuro-Clues from the Intersection of Disciplines," in Skov and Vartanian, *Neuroaesthetics*, 158, 164.

41. Oshin Vartanian, "Conscious Experience of Pleasure in Art," in Skov and Vartanian, *Neuroaesthetics*, 263.

42. Mari Tervaniemi, "Musical Sounds in the Human Brain," in Skov and Vartanian, *Neuroaesthetics*, 221–22. Also see Nikolus Steinbeis et al., "The Role of Harmonic Expectancy Violations in Musical Emotions: Evidence from Subjective, Physiological, and Neural Responses," *Journal of Cognitive Neuroscience* 18.8 (2006): 1380–93.

43. Daniel J. Levitin, *This is Your Brain on Music: The Science of a Human Obsession* (East Rutherford, NJ: Penguin, 2006), 74. For a concise survey of the neuroscientific research on music, see Isabelle Peretz, "The Nature of Music from a Biological Perspective," *Cognition* 100.1 (2006): 1–32.

44. Aniruddh D. Patel, *Music, Language, and the Brain* (Oxford: Oxford UP, 2008), 90.

45. Theodor W. Adorno, *Philosophy of New Music* (1949), trans. Robert Hullot-Kentor (Minneapolis: U of Minnesota P, 2006), 101, 102.

46. Jonah Lehrer, *Proust Was a Neuroscientist* (Boston: Houghton Mifflin, 2008), 125.

47. Sandra Trehub et al., "The Origins of Music Perception and Cognition: A Developmental Perspective," in *Perception and Cognition of Music*, ed. Irène Deliège and John Sloboda (East Sussex, UK: Psychology Press, 1997), 122.

48. See Bear, Connors, and Paradiso, *Neuroscience*, 544–46. These authors report as well that fMRI experiments on human subjects show "that regions of the brain dense with oxytocin and vasopressin receptors are activated when mothers look at photographs of their own children but not when they look at photographs of the children of their friends" (546). Patricia Churchland's argument that "morality originates in the neurobiology of attachment and bonding" is based, she explains, "on the idea that the oxytocin-vasopressin network can be modified to allow care to be extended to others beyond one's litter of juveniles." Churchland, *Braintrust: What Neuroscience Tells Us about Morality* (Princeton, NJ: Princeton UP, 2011), 71. The neurobiology of morality and other social phenomena entails more than brain chemistry, of course, as I explore in chapter 5 below.

49. Martin Skov and Oshin Vartanian, "Introduction: What is Neuroaesthetics?" in Skov and Vartanian, *Neuroaesthetics*, 3, 4.

50. For a report on imaging experiments that show how different areas of the brain respond to different kinds of musical dissonance, see Tervaniemi, "Musical Sounds in the Human Brain," 226.

51. William James, *The Principles of Psychology*, 2 vols. (1890; New York: Dover, 1950), 1:107.

52. Victor Shklovsky, "Art as Technique" (1917), in *Russian Formalist Criticism: Four Essays*, ed. Lee T. Lemon and Marion J. Reis (Lincoln: U of Nebraska P, 1965), 12, emphasis in original. Shklovsky quotes Leo Tolstoy's diary entry for 1 March 1897.

53. Hans Robert Jauss, "Literary History as a Challenge to Literary Theory," in *Toward an Aesthetic of Reception*, trans. Timothy Bahti (Minneapolis: U of Minnesota P, 1982), 25–26.

54. For a broad if, as its subtitle suggests, not always cautious survey of the field, see Daniel J. Linden, *The Compass of Pleasure: How Our Brains Make Fatty Foods, Orgasm, Exercise, Marijuana, Generosity, Vodka, Learning, and Gambling Feel So Good* (New York: Viking, 2011).

55. Irving Biederman and Edward A. Vessel, "Perceptual Pleasure and the Brain," *American Scientist* 94.3 (May–June 2006): 247–49.

56. Immanuel Kant, *Critique of Judgement* (1790), trans. James Creed Meredith, ed. Nicholas Walker (Oxford: Oxford UP, 2007), 49.

57. See Hans-Georg Gadamer, *Truth and Method* (1960), trans. Joel Weinsheimer and Donald G. Marshall, 2nd ed. (New York: Continuum, 1993), esp. 101–10; Wolfgang Iser, *Prospecting: From Reader Response to Literary Anthropology* (Baltimore: Johns Hopkins UP, 1989), 249–61; and Iser, *The Fictive and the Imaginary: Charting Literary Anthropology* (Baltimore: Johns Hopkins UP, 1993), 69–86, 247–80. Also see my book *Play and the Politics of Reading: The Social Uses of Modernist Form* (Ithaca, NY: Cornell UP, 2005), 2–41.

58. See two classic studies of the anthropological uses of play: Johan Huizinga, *Homo Ludens: A Study of the Play Element in Culture*, trans. anon. (1944; Boston: Beacon, 1950); and Roger Caillois, *Man, Play, and Games*, trans. Meyer Barash (1958; Urbana: U of Illinois P, 2001). Iser uses Caillois's theory to develop a taxonomy of kinds of aesthetic "play." See

Iser, *Fictive and the Imaginary*, 257–73. Also see Brian Boyd's important Darwinian case for the evolutionary value of aesthetic play in *On the Origin of Stories*. Working from a very different tradition (and apparently unaware of these theorists of play), Boyd reaches conclusions similar to mine: "We can define art as cognitive play with pattern," which "increases cognitive skills, repertoires, and sensitivities" and enhances our species' flexibility (15). On the convergence of evolutionary and neurobiological perspectives toward art, also see Martin Skov, "Neuroaesthetic Problems: A Framework for Neuroaesthetic Research," in Skov and Vartanian, *Neuroaesthetics*, 12.

59. See Boyd, *Origin of Stories*, 179. On how dopamine may promote the brain's response to novelty and thereby facilitate learning, see Pascale Waelti, Anthony Dickinson, and Wolfram Schultz, "Dopamine Responses Comply With Basic Assumptions of Formal Learning Theory," *Nature* 412 (5 July 2001): 43–48. On the role of what he calls "the dopamine pleasure circuit" in video-game play, see Linden, *Compass of Pleasure*, 144–47.

CHAPTER 3: The Neuroscience of the Hermeneutic Circle

1. For an account of this tradition and its relevance to contemporary literary theory, see my article "Hermeneutics" in *Blackwell Encyclopedia of Literary and Cultural Theory*, gen. ed. Michael Ryan, vol. 1, *Literary Theory from 1900 to 1966*, ed. Gregory Castle (Malden, MA: Wiley-Blackwell, 2011), 236–46.

2. See Wolfgang Iser, "The Reading Process: A Phenomenological Approach," in *The Implied Reader: Patterns of Communication in Prose Fiction from Bunyan to Beckett* (Baltimore: Johns Hopkins UP, 1974), 274–94; and Iser, *The Act of Reading: A Theory of Aesthetic Response* (Baltimore: Johns Hopkins UP, 1978), esp. 107–34.

3. For more on the interpretive disagreements this tale has generated, see the chapter "History, Epistemology, and the Example of *The Turn of the Screw*" in my *Conflicting Readings: Variety and Validity in Interpretation* (Chapel Hill: U of North Carolina P, 1990), 89–108.

4. See E. H. Gombrich, *Art and Illusion* (1960; Princeton, NJ: Princeton UP, 2000); and Semir Zeki, "The Neurology of Ambiguity," *Consciousness and Cognition* 13 (2004): 173–96. On how ambiguous figures reveal "the gestalt character of perception," also see Hubert L. Dreyfus, *What Computers Still Can't Do: A Critique of Artificial Reason* (Cambridge, MA: MIT P, 1992), 235–48: "The significance of the details and indeed their very look is determined by my perception of the whole" (238).

5. See Martin Heidegger, "Understanding and Interpretation," in *Being and Time* (1927), trans. John Macquarrie and Edward Robinson (New York: Harper & Row, 1962), 188–95.

6. For a masterful exploration of this aspect of narrative, see Roland Barthes's analysis of the "hermeneutic code" in *S/Z* (1970), trans. Richard Miller (New York: Hill & Wang, 1974).

7. See my *Conflicting Readings*, esp. 1–19.

8. The neuroscience of vision provides particularly interesting examples, as we shall see in detail below. See Zeki, "Neurology of Ambiguity"; Zeki, *Splendors and Miseries of the Brain: Love, Creativity, and the Quest for Human Happiness* (Malden, MA: Wiley-Blackwell, 2009), 59–97; and Margaret Livingstone, *Vision and Art: The Biology of Seeing* (New York: Abrams, 2002).

9. See Mark Bear, Barry W. Connors, and Michael A. Paradiso, *Neuroscience: Exploring the Brain*, 3rd ed. (Baltimore: Lippincott Williams & Wilkins, 2007), 340.

10. Livingstone, *Vision and Art*, 53.

11. For a history of the mirror metaphor in epistemology, see Richard Rorty, *Philosophy and the Mirror of Nature* (Princeton, NJ: Princeton UP, 1979). Rorty's influential pragmatic critique of this metaphor is consistent with the neuroscience of the hermeneutic circle. Also see Daniel C. Dennett's well-known refutation of the assumption that the mind is a "Cartesian Theater" in *Consciousness Explained* (New York: Little, Brown, 1991), esp. 101–38.

12. Semir Zeki, *Inner Vision: An Exploration of Art and the Brain* (Oxford: Oxford UP, 1999), 3.

13. Zoltán Jakab, "Opponent Processing, Linear Models, and the Veridicality of Color Perception," in *Cognition and the Brain: The Philosophy and Neuroscience Movement*, ed. Andrew Brook and Kathleen Akins (Cambridge: Cambridge UP, 2005), 373.

14. On the "pragmatic test" for validity in interpretation, see my *Conflicting Readings*, esp. 15–16. Also see Charles Sanders Peirce, "The Fixation of Belief" (1877), in *Philosophical Writings of Peirce*, ed. Justus Buchler (New York: Dover, 1955), 5–22.

15. Semir Zeki, "The Disunity of Consciousness," *Trends in Cognitive Science* 7.5 (May 2003): 214, 215.

16. My account here is based on Livingstone, *Vision and Art*, 40–45.

17. When the neuroscientist Gary Matthews, a former Stony Brook colleague of mine, read the manuscript of this book, he questioned the distinction Livingstone draws here between primates and other mammals: "X and Y cells (the equivalent of the primate P and M cells [the small, or *parvo*, and large, or *magno*, ganglion neurons]) were first discovered in cat retina, and X cells are found in many non-primate species. The commonly studied rodents (mice and rats) do not have a prominent X system (although it is there) because they are rod-dominant nocturnal animals. Ground squirrels and other diurnal rodents that are cone-dominant have a well-developed X cell system" and would consequently be able to make fine-grained visual distinctions. Personal communication.

18. Bear, Connors, and Paradiso, *Neuroscience*, 310.

19. Ferdinand de Saussure, *Course in General Linguistics*, ed. Charles Bally, Albert Sechehaye, and Albert Riedlinger, trans. Wade Baskin (New York: McGraw Hill, 1966), 120, emphasis in original.

20. Ellen Spolsky, "Making 'Quite Anew': Brain Modularity and Creativity," in *Introduction to Cognitive Cultural Studies*, ed. Lisa Zunshine (Baltimore: Johns Hopkins UP, 2010), 89.

21. See William James, *Principles of Psychology*, 2 vols. (1890; New York: Dover, 1950), 1:488.

22. For a description of the state of research, see Elkhonon Goldberg, *The New Executive Brain: Frontal Lobes in a Complex World* (New York: Oxford UP, 2009), 63–88. Different areas of the brain seem to be involved in these responses to varying degrees in a way that defies easy localization. This lack of specificity may reflect the early stages of the research, or more likely, as I will argue, it may suggest the reciprocal involvement of a variety of cortical areas in responding to novelty. Not surprisingly, perhaps, neuroaesthetic research has been particularly interested in these questions. For example, see Mari

Tervaniemi, "Musical Sounds in the Human Brain," in *Neuroaesthetics*, ed. Martin Skov and Oshin Vartanian (Amityville, NY: Baywood, 2009), 221–32. Because of the importance of surprise and the disruption of expectations in aesthetic experience, this may be an area in which collaborations between aesthetics and neuroscience would be especially useful and promising.

23. Francisco J. Varela, "The Specious Present: A Neurophenomenology of Time Consciousness," in *Naturalizing Phenomenology: Issues in Contemporary Phenomenology and Cognitive Science*, ed. Jean Petitot et al. (Stanford, CA: Stanford UP, 1999), 285, emphasis in original.

24. Gombrich, *Art and Illusion*, 5. Applying Gombrich to literary and linguistic cases of multiple meaning, Shlomith Rimmon defines ambiguity as the conjunction of two mutually exclusive but equally tenable possibilities of meaning. See her useful book *The Concept of Ambiguity—The Example of James* (Chicago: U of Chicago P, 1977). I offer a phenomenological analysis of ambiguity not as "a fact in the text" (as Rimmon refers to it) but as an event in the experience of reading in my analysis of Henry James's notoriously ambiguous novel *The Sacred Fount* in *The Challenge of Bewilderment: Understanding and Representation in James, Conrad, and Ford* (Ithaca, NY: Cornell UP, 1987), 29–62.

25. W. J. T. Mitchell, *Picture Theory: Essays on Verbal and Visual Representation* (Chicago: U of Chicago P, 1994), 49. See Ludwig Wittgenstein, *Philosophical Investigations*, trans. G. E. M. Anscombe (New York: Macmillan, 1958), 194–97.

26. See, e.g., Paolo Capotosto et al., "Frontoparietal Cortex Controls Spatial Attention through Modulations of Anticipatory Alpha Rhythms," *Journal of Neuroscience* 29.18 (6 May 2009): 5863–72. Bear, Connors, and Paradiso caution, however, against identifying attention with a precise cortical location: "Numerous cortical areas appear to be affected by attention. . . . Attention selectively increases brain activity, but the particular areas affected depend on the nature of the behavioral task performed." *Neuroscience*, 651. See Zeki, "Neurology of Ambiguity," 185–87, for a more extensive technical analysis of possible alternative interpretations of the brain scans of responses to bi-stable images.

27. Goldberg, *New Executive Brain*, 102.

28. György Buzsáki, *Rhythms of the Brain* (Oxford: Oxford UP, 2006), 29n.

29. See my *Conflicting Readings*, esp. 1–19, for an explanation of how the possibility of disagreement between interpreters about the presuppositions that should guide understanding can give rise to irreconcilable conflicts between hermeneutic methods. Also see Paul Ricoeur, *The Conflict of Interpretations*, ed. Don Ihde (Evanston, IL: Northwestern UP, 1974).

30. According to Bear, Connors, and Paradiso, "Precisely speaking, there may not be such a thing as normal color vision" because of the variations in how different people respond to the same wavelength on the spectrum: "In a group of males classified as normal trichromats [i.e., who respond to all three basic pigments], it was found that some require slightly more red than others to perceive yellow in a red-green mixture," and similarly, "if a group of people are asked to choose the wavelength of light that most appears green without being yellowish or bluish, there will be small variations in the choices." *Neuroscience*, 297.

31. Antonio Damasio, *Descartes' Error: Emotion, Reason, and the Human Brain* (New York: Putnam, 1994), 172–74. The "somatic marker" hypothesis suggests that emotions

are deeply involved in our orientation toward the future, a topic I analyze in more detail in the next chapter's discussion of the brain's temporality. For an application of Damasio's model to literary criticism, see Kay Young, *Imagining Minds: The Neuro-Aesthetics of Austen, Eliot, and Hardy* (Columbus: Ohio State UP, 2010).

32. See E. D. Hirsch Jr., *Validity in Interpretation* (New Haven, CT: Yale UP, 1967), esp. 209–35. For a critique of his position for reasons slightly different from those offered here, see my *Conflicting Readings*, 1–43. Although I wrote my book on interpretive conflict before encountering neuroscience, its arguments are entirely consistent with what I have since learned about the experimental evidence of brain functioning. Hirsch's claim that a determinate "meaning" must first be recognized before variable "significances" can be attached to it is at odds, however, with the experimental evidence for how the brain handles ambiguity.

33. See Gerald M. Reichler, "Perceptual Recognition as a Function of Meaningfulness of Stimulus Material," *Experimental Psychology* 81.2 (1967): 275–80. The examples that follow are taken from Stanislas Dehaene's discussion of this experiment in *Reading in the Brain: The Science and Evolution of a Human Invention* (New York: Viking, 2009), 48–49.

34. See D. E. Rumelhart and J. L. McClelland, "An Interactive Activation Model of Context Effects in Letter Perception, Part 2: The Contextual Enhancement Effect and Some Tests and Extensions of the Model," *Psychological Review* 89 (1982): 60–94.

35. José Morais, Luz Cary, et al., "Does Awareness of Speech as a Sequence of Phonemes Arise Spontaneously?," *Cognition* 7.4 (1979): 323–31.

36. José Morais, Paul Bertelson, et al., "Literacy Training and Speech Segmentation," *Cognition* 24.1 (1986): 45–64.

37. Dreyfus, *What Computers Still Can't Do*, 238.

38. Stephen Pinker, *The Language Instinct: How the Mind Creates Language* (1994; New York: Harper, 2007), 324. For his critique of linear "parsing," see 195–230. As he asserts, "Chomsky showed that word-chain devices are not just a bit suspicious; they are deeply, fundamentally, the wrong way to think about how human language works" (85).

39. See Iser, "Reading Process," 274–94; and Iser, *Act of Reading*, 163–231.

40. Richard J. Gerrig and Giovanna Egidi, "Cognitive Psychological Foundations of Narrative Experiences," in *Narrative Theory and the Cognitive Sciences*, ed. David Herman (Stanford, CA: Center for the Study of Language and Information Publications, 2003), 40–41. Also see Gerrig, *Experiencing Narrative Worlds: On the Psychological Activities of Reading* (New Haven, CT: Yale UP, 1993). In a useful correction of the perhaps overly rigid, Platonic implications of the term *schema*, Gerrig prefers in his recent work to describe reading as "memory-based processing" (personal communication). As he argues in a recent essay, "Readers' use of general knowledge" to make inferences about narrative gaps is "more fluid and idiosyncratic" than the application of preset schemas suggests. "Readers' Experiences of Narrative Gaps," *StoryWorlds* 2.1 (2010): 22. This does not change my point here, however, because making inferences based on memory is still a configurative process of consistency building.

41. Hans-Georg Gadamer, *Truth and Method* (1960), trans. Joel Weinsheimer and Donald G. Marshall, 2nd ed. (New York: Continuum, 1993), 268. For an exploration of how common psychological constructs for understanding persons as opposed to objects are laid bare by being invoked and questioned by literary works, see Lisa Zunshine,

Strange Concepts and the Stories They Make Possible (Baltimore: Johns Hopkins UP, 2008). For phenomenologically oriented analyses of such experiences, see Iser's analyses of a range of literary works, from Bunyan's *Pilgrim's Progress* to the modernist novels of Joyce and Beckett, in *The Implied Reader.* Also see my *Challenge of Bewilderment* and my *Play and the Politics of Reading: The Social Uses of Modernist Form* (Ithaca, NY: Cornell UP, 2005).

42. See the chapter "The Cognitive Powers of Metaphor" in my *Conflicting Readings,* 67–88. The seminal texts of interaction theory are Nelson Goodman, *Languages of Art,* 2nd ed. (Indianapolis: Hackett, 1976); and Paul Ricoeur, *The Rule of Metaphor,* trans. Robert Czerny (Toronto: U of Toronto P, 1977). The work of George Lakoff and Mark Johnson, which is often cited by contemporary cognitive literary critics, carries on this tradition. See their *Metaphors We Live By* (Chicago: U of Chicago P, 1980). Lakoff and Johnson apply their ideas about metaphor to cognitive science in *Philosophy in the Flesh: The Embodied Mind and Its Challenge to Western Thought* (New York: Basic Books, 1999). Also see David Miall's notion of a "defamiliarization-reconceptualization cycle" in literary reading. "Neuroaesthetics of Literary Reading," in Skov and Vartanian, *Neuroaesthetics,* 235–36.

43. Friedrich Nietzsche, "Über Wahrheit und Lüge im aussermoralischen Sinn" [On Truth and Lie in an Extra-Moral Sense] (1873), in *Werke in drei Bänden,* ed. Karl Schlechta (Munich: Hanser, 1977), 3:313.

44. Mark Turner, "The Cognitive Study of Art, Literature, and Language," *Poetics Today* 23.1 (Spring 2002): 10. For a fuller statement of his position, see Turner, *The Literary Mind: The Origins of Thought and Language* (New York: Oxford UP, 1996).

45. Hence the title of my book *The Challenge of Bewilderment,* a study of the role of perceptual blockages in prompting epistemological self-consciousness in the experience of reading and in the dramatized lives of characters in works of Henry James, Joseph Conrad, and Ford Madox Ford. The art historian Barbara Maria Stafford similarly observes: "The brain's 'binding' capacity, no less than gene splicing or blood chemistry, can be externalized for scrutiny precisely in those aesthetic situations when it abruptly juxtaposes or tightly links variegated experiences." "The Combinatorial Aesthetics of Neurobiology," in *Aesthetic Subjects,* ed. Pamela R. Matthews and David McWhirter (Minneapolis: U of Minnesota P, 2003), 253.

46. See Jonah Lehrer, *Proust Was a Neuroscientist* (Boston: Houghton Mifflin, 2008), which proposes a series of parallels between the findings of contemporary neuroscience and works by artists such as Walt Whitman, George Eliot, Igor Stravinsky, and Gertrude Stein. Zeki similarly argues that "artists are neurologists, studying the organisation of the visual brain with techniques unique to them." *Inner Vision,* 202.

CHAPTER 4: The Temporality of Reading and the Decentered Brain

1. Augustine, *Confessions,* bk. 11, quoted by Edmund Husserl in *The Phenomenology of Internal Time Consciousness* (1928), ed. Martin Heidegger, trans. James S. Churchill (Bloomington: Indiana UP, 1964), 21: "Si nemo a me quaerat, scio, si quaerenti explicare velim, nescio."

2. William James, *Principles of Psychology,* 2 vols. (1890; New York: Dover, 1950), 1:609.

3. Maurice Merleau-Ponty, *Phenomenology of Perception* (1945), trans. Colin Smith (London: Routledge & Kegan Paul, 1962), 418.

4. Shaun Gallagher and Dan Zahavi, *The Phenomenological Mind: An Introduction to Philosophy of Mind and Cognitive Science* (New York: Routledge, 2008), 75.

5. See Tim Van Gelder, "Wooden Iron? Husserlian Phenomenology Meets Cognitive Science," in *Naturalizing Phenomenology: Issues in Contemporary Phenomenology and Cognitive Science*, ed. Jean Petitot et al. (Stanford, CA: Stanford UP, 1999), 245–65.

6. Wolfgang Iser, *The Implied Reader: Patterns of Communication in Prose Fiction from Bunyan to Beckett* (Baltimore: Johns Hopkins UP, 1974), 280.

7. Wolfgang Iser, *The Act of Reading: A Theory of Aesthetic Response* (Baltimore: Johns Hopkins UP, 1978), 128, 132, 148.

8. Bernard J. Baars and Nicole M. Gage, *Cognition, Brain, and Consciousness*, 2nd ed. (Amsterdam: Elsevier, 2010), 177. See also Semir Zeki, "Cerebral Akinetopsia (Visual Motion Blindness)," *Brain* 114 (1991): 811–24.

9. See Husserl, *Phenomenology of Internal Time Consciousness*, esp. 48–63. Also see Dan Zahavi, *Subjectivity and Selfhood: Investigating the First-Person Perspective* (Cambridge, MA: MIT P, 2008), 49–72; Evan Thompson, *Mind in Life: Biology, Phenomenology, and the Sciences of Mind* (Cambridge, MA: Harvard UP, 2007), 312–59; and my article "Intentionality and Horizon," in *Blackwell Encyclopedia of Literary and Cultural Theory*, gen. ed. Michael Ryan, vol. 1, *Literary Theory from 1900 to 1966*, ed. Gregory Castle (Malden, MA: Wiley-Blackwell, 2011), 263–68.

10. Jean-Michel Roy et al., "Beyond the Gap," in Petitot et al., *Naturalizing Phenomenology*, 27.

11. Francisco J. Varela, "The Specious Present: A Neurophenomenology of Time Consciousness," in Petitot et al., *Naturalizing Phenomenology*, 296.

12. See Aniruddh D. Patel, *Music, Language, and the Brain* (Oxford: Oxford UP, 2008), 174, 228, 238, 268.

13. Lawrence M. Zbikowski, "The Cognitive Tango," in *The Artful Mind: Cognitive Science and the Riddle of Human Creativity*, ed. Mark Turner (Oxford: Oxford UP, 2006), 128.

14. See Daniel J. Levitin's entertaining and enlightening book on the neuroscience of music, *This is Your Brain on Music: The Science of a Human Obsession* (East Rutherford, NJ: Penguin, 2006), 124–27.

15. See Mark Bear, Barry W. Connors, and Michael A. Paradiso, *Neuroscience: Exploring the Brain*, 3rd ed. (Baltimore: Lippincott Williams & Wilkins, 2007), 620–25.

16. That is, of course, Jane Austen's *Pride and Prejudice*, Leo Tolstoy's *Anna Karenina*, and Franz Kafka's *The Trial*.

17. Iser, *Act of Reading*, 108, 111. Also see Roman Ingarden's chapter "Temporal Perspective in the Concretization of the Literary Work of Art," in *The Cognition of the Literary Work of Art* (1937), trans. Ruth Ann Crowley and Kenneth R. Olson (Evanston, IL: Northwestern UP, 1973), 94–145.

18. See the chapter "Varieties of the Cognition of the Literary Work of Art," in Ingarden, *Cognition of the Literary Work of Art*, 168–331.

19. Varela, "Specious Present," 272–73. This article from the late 1990s is the *locus classicus* of neurophenomenological investigations of time. For explanations of Varela's sometimes dense but always perspicacious analyses, see Thompson, *Mind in Life*, esp.

329–38; and Gallagher and Zahavi, *Phenomenological Mind*, 80–82. Also see the early and still important book by Francisco J. Varela, Evan Thompson, and Eleanor Rosch, *The Embodied Mind: Cognitive Science and Human Experience* (Cambridge, MA: MIT P, 1991).

20. See Varela, Thompson, and Rosch, *Embodied Mind*, 73–74.

21. See Bear, Connors, and Paradiso, *Neuroscience*, 369–71. Also see James A. Simmons, "A View of the World Through the Bat's Ear: The Formation of Acoustic Images in Echolation," *Cognition* 33.1 (1989): 155–99.

22. Semir Zeki, *Splendors and Miseries of the Brain: Love, Creativity, and the Quest for Human Happiness* (Malden, MA: Wiley-Blackwell, 2009), 37. As Zeki explains, "Color and motion have different temporal requirements. With color, signals coming from many different parts of the field have to be compared simultaneously. . . . With motion, signals coming from different points successively in time must be compared" (37).

23. Semir Zeki, "The Disunity of Consciousness," *Trends in Cognitive Science* 7.5 (May 2003): 215.

24. Varela, "Specious Present," 273–74. See Thompson, *Mind in Life*, 331–32; and Gallagher and Zahavi, *Phenomenological Mind*, 81–82.

25. The classic texts on futurity in the existential tradition are Martin Heidegger, *Being and Time* (1927), trans. John Macquarrie and Edward Robinson (New York: Harper & Row, 1962), esp. 279–311 and 383–423; and Jean-Paul Sartre, *Being and Nothingness* (1945), trans. Hazel E. Barnes (New York: Washington Square, 1966), 159–237.

26. Antonio Damasio, *The Feeling of What Happens: Body and Emotion in the Making of Consciousness* (New York: Harcourt Brace, 1999), 127.

27. William James quotes Kierkegaard in *Pragmatism: A New Name for Some Old Ways of Thinking* (1907; Cambridge, MA: Harvard UP, 1978): "We live forwards, a Danish thinker has said, but we understand backwards. The present sheds a backward light on the world's previous processes" (107). The anonymous editors of *Pragmatism* attribute this reference to *The Journals of Søren Kierkegaard*, ed. and trans. Alexander Dru (London: Oxford UP, 1938), 127.

28. See Seymour Chatman, *Story and Discourse: Narrative Structure in Fiction and Film* (Ithaca, NY: Cornell UP, 1978).

29. See Frank Kermode, *The Sense of an Ending: Studies in the Theory of Fiction* (Oxford: Oxford UP, 1967).

30. See Gérard Genette, *Narrative Discourse: An Essay in Method*, trans. Jane E. Levin (Ithaca, NY: Cornell UP, 1980).

31. Ford Madox Ford, *Joseph Conrad: A Personal Remembrance* (Boston: Little, Brown, 1924), 136. On the fraught relationship between these two writers, see Thomas Moser, *The Life in the Fiction of Ford Madox Ford* (Princeton, NJ: Princeton UP, 1981).

32. For more extensive analyses of these experiments and their epistemological consequences, see my book *The Challenge of Bewilderment: Understanding and Representation in James, Conrad, and Ford* (Ithaca, NY: Cornell UP, 1987), esp. 1–25, 109–48, 189–224.

33. The explanation that follows is based on the chapter "The Action Potential" in Bear, Connors, and Paradiso, *Neuroscience*, 75–100.

34. I owe this analogy to the neuroscientist Jim McIlwain, a fellow Brown faculty member: "Action potentials do not travel down axons in the same way that electrons move through a copper wire. Charges are transferred only at the current site of the action

potential, which then triggers an action potential at the adjacent excitable site and so on. It is a continuous process, more like flame moving along a fire-cracker's fuse than like electrons flowing down a wire." Personal communication.

35. I owe this anecdote to a personal communication from the Stony Brook neuroscientist Gary Matthews.

36. Lively and highly informative, if somewhat technical in parts, the authoritative text on brain waves is György Buzsáki, *Rhythms of the Brain* (Oxford: Oxford UP, 2006).

37. My account of the neuroscience of smell is based on "Spatial and Temporal Representations of Olfactory Information" in Bear, Connors, and Paradiso, *Neuroscience*, 272–74.

38. See G. Laurent, "Olfactory Network Dynamics and the Coding of Multidimensional Signals," *Nature Reviews/Neuroscience* 3 (2002): 884–95.

39. Antonio Damasio, *Self Comes to Mind: Constructing the Conscious Brain* (New York: Pantheon, 2010), 87.

40. Thompson, *Mind in Life*, 330, 332. For a more technical explanation, see Buzsáki, *Rhythms of the Brain*, esp. 136–74.

41. For useful maps of the brain regions that interact in music processing, see Levitin, *This is Your Brain on Music*, 270–71.

42. Also see Thompson, *Mind in Life*, 329–49; and Gallagher and Zahavi, *Phenomenological Mind*, 80–82.

43. I owe this analogy, as well as my account in what follows of the brain's multiple wavelengths, to Baars and Gage, *Cognition, Brain, and Consciousness*, 101–8, 244–55.

44. Colin Martindale, *Cognitive Psychology: A Neural-Network Approach* (Pacific Grove, CA: Brooks/Cole, 1991), 13.

45. Baars and Gage credit as their source here G. Tononi, "An Information Integration Theory of Consciousness," *BMC Neuroscience* 5 (2004): 42.

46. See Roman Ingarden's section "The Literary Work of Art and the Polyphonic Harmony of its Aesthetic Value Qualities" in *The Literary Work of Art* (1931), trans. George G. Grabowicz (Evanston, IL: Northwestern UP, 1973), 369–73. The emphasis here on *polyphonic* is mine.

47. See Victor Shklovsky, "Art as Technique" (1917), in *Russian Formalist Criticism: Four Essays*, ed. Lee T. Lemon and Marion J. Reis (Lincoln: U of Nebraska P, 1965), 3–24; and Hans Robert Jauss, "Literary History as a Challenge to Literary Theory," in *Toward an Aesthetic of Reception*, trans. Timothy Bahti (Minneapolis: U of Minnesota P, 1982), 3–45.

48. David S. Miall, *Literary Reading: Empirical and Theoretical Studies* (New York: Peter Lang, 2006), 3. Also see Lisa Zunshine, *Strange Concepts and the Stories They Make Possible* (Baltimore: Johns Hopkins UP, 2009).

49. Brian Boyd, *On the Origin of Stories: Evolution, Cognition, and Fiction* (Cambridge, MA: Harvard UP, 2009), 135, emphasis in original.

50. Alva Noë, *Out of Our Heads: Why You are Not Your Brain, and Other Lessons from the Biology of Consciousness* (New York: Hill & Wang, 2009), 100. The source of the experimental evidence he cites is John Milton et al., "The Mind of Expert Motor Performance is Cool and Focused," *NeuroImage* 35 (2007): 804–13.

51. See Hans-Georg Gadamer, *Truth and Method* (1960), trans. Joel Weinsheimer and Donald G. Marshall, 2nd ed. (New York: Continuum, 1993), 101–10.

52. Noë's excellent chapter "Habits" (*Out of Our Heads*, 99–128) provides a twenty-

first-century reformulation of the pragmatist arguments classically laid out by William James in his eloquent chapter "Habit," in *Principles of Psychology*, 1:104–27.

53. For a detailed summary of this experiment, see Bear, Connors, and Paradiso, *Neuroscience*, 765–71. Unfortunately, Kandel does not discuss the aesthetic implications of this work in his interesting recent book on neuroscience and art, *The Age of Insight: The Quest to Understand the Unconscious in Art, Mind, and Brain from Vienna 1900 to the Present* (New York: Random House, 2012).

54. See Thomas Nagel, "What Is It Like to Be a Bat?," *Philosophical Review* 83 (1974): 435–50.

55. See I. Gauthier et al., "Expertise for Cars and Birds Recruits Brain Areas Involved in Face Recognition," *Nature Neuroscience* 3 (2000): 191–97. For a discussion of this experiment, see Bear, Connors, and Paradiso, *Neuroscience*, 735–36.

56. This is the neurobiological explanation of how an interpreter's learned assumptions about literature, language, and life will inform his or her habitual practices of generating hypotheses about textual meaning, a relation I describe as an interaction between two levels of belief in understanding in *Conflicting Readings: Variety and Validity in Interpretation* (Chapel Hill: U of North Carolina P, 1990), esp. 2–12.

57. Robert Darnton, "What is the History of Books?" (1982) in *The Book History Reader*, ed. David Finkelstein and Alistair McCleery (New York: Routledge, 2006), 20.

58. Philip J. Ethington similarly argues for distinguishing on neuroscientific grounds between different historical time scales in his interesting essay "Sociovisual Perspective: Vision and the Forms of the Human Past," in *A Field Guide to a New Meta-Field: Bridging the Humanities-Neurosciences Divide*, ed. Barbara Maria Stafford (Chicago: U of Chicago P, 2011), 123–52.

59. Walter Ong, "Orality and Literacy: Writing Restructures Consciousness" (1997), in Finkelstein and McCleery, *Book History Reader*, 134.

60. Stanislas Dehaene, *Reading in the Brain: The Science and Evolution of a Human Invention* (New York: Viking, 2009), 4.

61. Jauss, "Literary History as a Challenge to Literary Theory," 41, 45; Nicholas Dames, *The Physiology of the Novel: Reading, Neural Science, and the Form of Victorian Fiction* (New York: Oxford UP, 2007), 7, 10. Attempting to negotiate a course between these poles, Suzanne Keen provides a judicious assessment of contemporary arguments about "the question of what a habit of novel reading does to the moral imagination of the immersed reader" in her interesting book *Empathy and the Novel* (New York: Oxford UP, 2007), esp. vii–xxv. I take up the issue of empathy in the next chapter's discussion of mirror neurons. Here my concern is with the role of habituation and dehabituation in learning.

62. Traumatic experiences would seem to be an exception to this rule, although the neuroscience of trauma is still a developing area of research. It is safe to say, however, that a reading experience is unlikely to have a traumatic effect, no matter how powerful its emotional impact. On traumatic experiences, see esp. Judith L. Herman, *Trauma and Recovery* (New York: Basic Books, 1992); and Robert C. Scaer, *The Trauma Spectrum: Hidden Wounds and Human Resiliency* (New York: Norton, 2005).

63. Damasio, *Self Comes to Mind*, 22. Also see his important earlier books on the neuroscience of emotions, esp. *Descartes' Error: Emotion, Reason, and the Human Brain* (New York: Putnam, 1994) and *The Feeling of What Happens* (see n. 26 above).

64. G. Gabrielle Starr, "Multisensory Imagery," in *Introduction to Cognitive Cultural Studies*, ed. Lisa Zunshine (Baltimore: Johns Hopkins UP, 2010), 280.

65. Thomas F Münte et al., "The Musician's Brain as a Model of Neuroplasticity," *Nature Reviews/Neuroscience* 3 (June 2002): 476.

66. See Olaf Hauk and Friedman Pulvermüller, "Neurophysiological Distinction of Action Words in the Fronto-Central Cortex," *Human Brain Mapping* 21.3 (2004): 191–201; and Véronique Boulenger et al., "Cross-Talk Between Language Processes and Overt Motor Behavior in the First 200 msec of Processing," *Journal of Cognitive Neuroscience* 18.10 (2006): 1607–15. Also see the discussion of mirror neurons in chapter 5 below.

67. See Roel W. Willems et al., "Body-Specific Representations of Action Verbs: Neural Evidence from Right- and Left-Handers," *Psychological Science* 21.1 (2010): 67–74. Willems found that action-related verbs triggered areas of the premotor cortex in the hemisphere contralateral to the subject's dominant hand: "Implicit mental simulation during language processing is body specific: Right- and left-handers, who perform actions differently, use correspondingly different areas of the brain for representing action verb meanings" (67).

68. See Damasio, *Descartes' Error*, esp. 205–22.

69. See Heidegger, *Being and Time*, 172–79. Thompson justly criticizes Heidegger's "account of mood and attunement" as "strangely disembodied" and consequently "unsatisfying" "from a neurophenomenological point of view." *Mind in Life*, 379. This deficiency reflects Heidegger's concern first and foremost with questions of fundamental ontology, for which, in his view, the ontic particularities of embodied existence matter only as indirect evidence of Being. This emphasis can be shifted, however, as it was by Merleau-Ponty, who took Heidegger's descriptions of the structures of existence as a guide for his explorations of the lived experience of embodiment in *Phenomenology of Perception*. Other existential phenomenologists (including Thompson) have followed Merleau-Ponty's lead.

70. See Gallagher and Zahavi, *Phenomenological Mind*, 33–38.

71. See David S. Miall, "Neuroaesthetics of Literary Reading," in *Neuroaesthetics*, ed. Martin Skov and Oshin Vartanian (Amityville, NY: Baywood, 2009), 237–40.

72. The role of expectations in music has been studied particularly thoroughly. For a useful overview, see Levitin, *This is Your Brain on Music*, 111–31.

73. Alvin I. Goldman, *Simulating Minds: The Philosophy, Psychology, and Neuroscience of Mindreading* (Oxford: Oxford UP, 2006), 161–62. The original accounts of these experiments are J. A. Bargh et al., "Automaticity of Social Behavior: Direct Effects of Trait Construct and Stereotype Activation on Action," *Journal of Personality and Social Psychology* 71 (1996): 230–44; and A. Dijksterhuis and A. van Knippenberg, "Behavioral Indecision: Effects of Self Focus on Automatic Behavior," *Social Cognition* 18 (2000): 55–74.

74. Starr, "Multisensory Imagery," 285, 331. See Jelena Djordjevic, R. J. Zatorre, and M. Jones-Gotman, "Effects of Perceived and Imagined Odors on Taste Detection," *Chemical Senses* 29 (2004): 199–208; and Djordjevic, Zatorre, et al., "The Mind's Nose: Effects of Odor and Visual Imagery on Odor Detection," *Psychological Science* 15.3 (2004): 143–48.

75. Paul Bloom, *How Pleasure Works* (New York: Norton, 2010), 192. The classic study of the effects on the viewer of representations of violence is Albert Bandura et al., "Imitation of Film-Mediated Aggressive Models," *Journal of Abnormal and Social Psychology* 66.1

(1963): 3–11. Although the capacity of depictions of violence to elicit aggressive feelings is well documented, the extent to which we necessarily imitate what we see is a controversial and complicated question, as I show in more detail in the next chapter.

76. Roy et al., "Beyond the Gap" (see n. 10 above), 35. On the *Körper-Leib* distinction, also see Thompson, *Mind in Life*, 235–37; and Zahavi, *Subjectivity and Selfhood*, 205–6. In chapter 5 I analyze in detail the neurological bases of self-other relations.

77. Gallagher and Zahavi, *Phenomenological Mind*, 202. "From a purely descriptive point of view," Gallagher and Zahavi argue, "there is nothing new in the analyses offered by Damasio. We are dealing with a reformulation of ideas already found in classical phenomenology" (203).

78. John R. Searle, "Consciousness," *Annual Review of Neuroscience* 23 (2000): 557, 562.

79. Jean-Luc Petit, "Constitution by Movement: Husserl in Light of Recent Neurobiological Findings," in Petitot et al., *Naturalizing Phenomenology*, 222.

CHAPTER 5: The Social Brain and the Paradox of the Alter Ego

1. Patricia S. Churchland, *Braintrust: What Neuroscience Tells Us about Morality* (Princeton, NJ: Princeton UP, 2011), 9.

2. Marco Iacoboni, *Mirroring People: The Science of Empathy and How We Connect with Others* (New York: Farrar, Straus & Giroux, 2008), 152.

3. Evan Thompson, *Mind in Life: Biology, Phenomenology, and the Sciences of Mind* (Cambridge, MA: Harvard UP, 2007), 403.

4. Merlin Donald, *A Mind So Rare: The Evolution of Human Consciousness* (New York: Norton, 2001), 212.

5. See Simon Baron-Cohen, *Mindblindness: An Essay on Autism and Theory of Mind* (Cambridge, MA: MIT P, 1995); and Peter Carruthers and Peter K. Smith, eds., *Theories of Theories of Mind* (Cambridge: Cambridge UP, 1996).

6. See R. M. Gordon, "Folk Psychology as Simulation," *Mind and Language* 1.2 (1986): 158–71; M. Davies and T. Stone, eds., *Mental Simulation: Evaluations and Applications* (Oxford: Blackwell, 1995); and Alvin I. Goldman, *Simulating Minds: The Philosophy, Psychology, and Neuroscience of Mindreading* (Oxford: Oxford UP, 2006). A good summary of theory theory and simulation theory, along with further bibliographical references, can be found in Shaun Gallagher, *How the Body Shapes the Mind* (Oxford: Clarendon, 2005), 206–8.

7. Shaun Gallagher and Dan Zahavi, *The Phenomenological Mind: An Introduction to Philosophy of Mind and Cognitive Science* (New York: Routledge, 2008), 175.

8. Christian Keysers and Valeria Gazzola, "Integrating Simulation Theory and Theory of Mind: From Self to Social Cognition," *Trends in Cognitive Sciences* 11.5 (March 2007): 194.

9. The classic paper reporting this discovery is Vittorio Gallese et al., "Action Recognition in the Premotor Cortex," *Brain* 119.2 (1996): 593–609. Also see Giacomo Rizzolatti et al., "Neurophysiological Mechanisms Underlying the Understanding and Imitation of Action," *Nature Reviews/Neuroscience* 2 (September 2001): 661–70; Giacomo Rizzolatti and Laila Craighero, "The Mirror Neuron System," *Annual Review of Neuroscience* 27 (2004):

169–92; and Giacomo Rizzolatti and Maddalena Fabbri-Destro, "Mirror Neurons: From Discovery to Autism," *Experimental Brain Research* 200 (2010): 223–37. Rizzolatti notes in this last paper that his first report of the discovery of mirror neurons was rejected by *Nature* for its "lack of general interest" (223)! This research is lucidly summarized in Giacomo Rizzolatti and Corrado Sinigaglia, *Mirrors in the Brain: How Our Minds Share Actions and Emotions*, trans. Frances Anderson (Oxford: Oxford UP, 2008).

10. Vittorio Gallese and Alvin Goldman, "Mirror Neurons and the Simulation Theory of Mind-Reading," *Trends in Cognitive Science* 2.12 (December 1998): 498. Also see Vittorio Gallese, "The 'Shared Manifold' Hypothesis: From Mirror Neurons to Empathy," *Journal of Consciousness Studies* 8.5–7 (2001): 33–50; and Gallese, "The Manifold Nature of Interpersonal Relations: The Quest for a Common Mechanism," *Philosophical Transactions of the Royal Society London B* 358 (2003): 517–28.

11. V. S. Ramachandran, *The Tell-Tale Brain: A Neuroscientist's Quest for What Makes Us Human* (New York: Norton, 2011), 121.

12. See G. Hickok, "Eight Problems for the Mirror Neuron Theory of Action Understanding in Monkeys and Humans," *Journal of Cognitive Neuroscience* 27.7 (2009): 1229–43. Churchland also subjects the mirror-neuron evidence to what she calls "tough tire-kicking" in *Braintrust*, 118–62, but does not reject it entirely. Ramachandran offers a concise reply to the most frequent skeptical arguments in *Tell-Tale Brain*, 312–13. Also see Christian Keysers, "Mirror Neurons," *Current Biology* 19.21 (2009): R971–73.

13. On the thought experiment in which the brain is imagined to exist in a vat and the question is then asked, "Would you (or your brain) notice any difference?," see Thompson, *Mind in Life*, 240–42, quotation from 240. Thompson answers: "As conscious subjects we are not brains in cranial vats; we are neurally enlivened beings in the world" (242). That world is social, and the question is how the brain's capacities allow us to navigate it.

14. I take this phrase from Maurice Merleau-Ponty, *Phenomenology of Perception* (1945), trans. Colin Smith (London: Routledge & Kegan Paul, 1962), xii.

15. Edmund Husserl, *Cartesian Meditations* (1929), trans. Dorion Cairns (The Hague: Martinus Nijhoff, 1970), 91.

16. The term *primary intersubjectivity* is Gallagher's in *How the Body Shapes the Mind*, 225–28. For a survey of phenomenological thinking about self and other, see Dan Zahavi, *Subjectivity and Selfhood: Investigating the First-Person Perspective* (Cambridge, MA: MIT P, 2008), 147–77.

17. From Friedrich Nietzsche, "Über Wahrheit und Lüge im aussermoralischen Sinn" [On Truth and Lie in an Extra-Moral Sense] (1873), in *Werke in drei Bänden*, ed. Karl Schlechta (Munich: Hanser, 1977), 3:309–22.

18. Georges Poulet, "Phenomenology of Reading" (1969) in *Critical Theory Since Plato*, ed. Hazard Adams (New York: Harcourt Brace Jovanovich, 1971), 1214.

19. Wolfgang Iser, *The Act of Reading: A Theory of Aesthetic Response* (Baltimore: Johns Hopkins UP, 1978), 166.

20. Henry James, "Alphonse Daudet" (1883), in *Theory of Fiction: Henry James*, ed. James E. Miller Jr. (Lincoln: U of Nebraska P, 1972), 93. On James's understanding of the paradox of the alter ego, see "Self and Other: Conflict versus Care in *The Golden Bowl*," in my book *The Phenomenology of Henry James* (Chapel Hill: U of North Carolina P, 1983), 136–86.

21. Jean-Paul Sartre, *What is Literature?* (1947), trans. Bernard Frechtman (New York: Harper & Row, 1965), 39.

22. Wolfgang Iser, *The Implied Reader: Patterns of Communication in Prose Fiction from Bunyan to Beckett* (Baltimore: Johns Hopkins UP, 1974), 293.

23. The findings described here are based on the reports in Rizzolatti and Sinigaglia, *Mirrors in the Brain*, 79–114; and Iacoboni, *Mirroring People*, 3–78.

24. Rizzolatti et al., "Neurophysiological Mechanisms Underlying the Understanding and Imitation of Action," 661.

25. This is what Gallese calls the "shared manifold" of intersubjectivity, "a basic level of our interpersonal interactions that does not make explicit use of propositional attitudes" but undergirds our "basic capacity to model the behavior of other individuals by employing the same resources used to model our own behavior" ("Manifold Nature of Interpersonal Relations," 525). Also see Gallagher, *How the Body Shapes the Mind*, 220–28.

26. See Leo Fogassi et al., "Parietal Lobe: From Action Organization to Intention Understanding," *Science* 308 (2005): 662–67. Iacoboni summarizes this experiment in *Mirroring People*, 31–33.

27. See Maria Alessandra Umiltà et al., "I Know What You Are Doing: A Neurophysiological Study," *Neuron* 31 (2001): 155–65. Also see Iacoboni's explanation in *Mirroring People*, 28–29.

28. Pierre Jacob similarly argues that mirror neurons alone cannot account for what he calls "embedded intentions." "It is highly questionable whether by mentally rehearsing an agent's observed movements, an observer could represent another's underlying intention," because "one and the same motor intention . . . could be at the service of different (even incompatible) prior intentions," as when someone flips off a light switch to go to bed, or to save electricity, or to improve viewing conditions for watching a movie, or for some other reason. "What Do Mirror Neurons Contribute to Human Social Cognition?," *Mind and Language* 23.2 (April 2008): 191, 206.

29. Rizzolatti and Fabbri-Destro, "Mirror Neurons: From Discovery to Autism," 230.

30. Report from William George Clark in *Fraser's* magazine (December 1849), quoted in Iser, *Implied Reader*, 291–92. This is the kind of phenomenon that Norman Holland's "neuropsychoanalysis" foregrounds; see his *Literature and the Brain* (Gainesville, FL: Psy-Art Foundation, 2009), esp. 40–124.

31. Ludwig Wittgenstein, *Remarks on the Philosophy of Psychology II* (Oxford: Blackwell, 1980), paragraph 570, emphasis in original.

32. See J. A. C. J. Bastiaansen et al., "Evidence for Mirror Systems in Emotions," *Philosophical Transactions of the Royal Society B* 364 (2009): 2396–97.

33. See Goldman, *Simulating Minds*, chap. 6, "Simulation in Low-Level Mindreading" (113–46), and chap. 7, "High-Level Simulational Mindreading" (147–91).

34. Blakey Vermeule, *Why Do We Care about Literary Characters?* (Baltimore: Johns Hopkins UP, 2010), xii. As Vermeule notes, the classic study of this topic is Patricia Meyer Spacks, *Gossip* (New York: Knopf, 1985).

35. See the sections entitled "Idle Talk" and "Curiosity" in Martin Heidegger, *Being and Time* (1927), trans. John Macquarrie and Edward Robinson (New York: Harper & Row, 1962), 211–17.

36. The narratological literature is vast and too large to encompass in a footnote. Re-

cently, see esp. David Herman, *Story Logic: Problems and Possibilities of Narrative* (Lincoln: U of Nebraska P, 2002), which proposes the term *storyworld* to mediate between the discourses of narrative and the cognitive acts of writers and readers through which textual worlds are constructed. On the implications of a writer's choice of a mode of narration, see Jonathan Culler's thoughtful recent essay "Omniscience" in his volume *The Literary in Theory* (Stanford, CA: Stanford UP, 2007), 183–201. On the relation of the narrator to the "narratee," see esp. Ross Chambers, *Room for Maneuver: Reading (the) Oppositional (in) Narrative* (Chicago: U of Chicago P, 1991). The classic text and still unparalleled source on all of these matters is, of course, Wayne Booth, *The Rhetoric of Fiction* (Chicago: U of Chicago P, 1961).

37. The standard narratological reference on this topic is still Dorrit Cohn, *Transparent Minds: Narrative Modes for Representing Consciousness in Fiction* (Princeton, NJ: Princeton UP, 1978). I analyze the experiential and epistemological implications of focalization in greater detail in *The Challenge of Bewilderment: Understanding and Representation in James, Conrad, and Ford* (Ithaca, NY: Cornell UP, 1987).

38. Grit Hein and Tania Singer, "I Feel How You Feel But Not Always: The Empathic Brain and its Modulation," *Current Opinion in Neurobiology* 18 (2008): 153.

39. Chris D. Frith and Uta Frith, "The Neural Basis of Mentalizing," *Neuron* 50 (2006): 531–32.

40. Bastiaansen et al., "Evidence for Mirror Systems in Emotions," 2398.

41. Lisa Zunshine, *Why We Read Fiction: Theory of Mind and the Novel* (Columbus: Ohio State UP, 2006), 164.

42. Brian Boyd, "Fiction and Theory of Mind," *Philosophy and Literature* 30 (October 2006): 590–600; Lisa Zunshine, "Fiction and Theory of Mind: An Exchange," *Philosophy and Literature* 31 (April 2007): 190, emphasis in original.

43. See I. Dinstein et al., "A Mirror Up to Nature," *Current Biology* 18 (2008): R13–18.

44. Keysers, "Mirror Neurons," R971–73. Keysers asks: "If there are mirror neurons, shouldn't all experiments find evidence for them? A basic power analysis falsifies this intuition: using the typical voxelwise thresholds of p< 0.001 in fMRI, one would expect even large effect sizes to remain undetected over 50% of the time" (R972). A voxel with a volume of 0.001 mm^3 is a large enough black box to house the speculations of both advocates and opponents of the mirror-neuron theory.

45. Christian Keysers, "Social Neuroscience: Mirror Neurons Recorded in Humans," *Current Biology* 20 (2010): R353–54. The experimental findings were reported by the neurosurgeons Roy Mukamel and Itzhak Fried, who teamed up with the neuroscientists Arne D. Ekstrom and Jonas Kaplan, from Iacoboni's group, in their paper "Single-Neuron Responses in Humans During Execution and Observation of Actions," *Current Biology* 20 (2010): 750–56.

46. For a summary of this evidence, see the chapter "Mirror Neurons in Humans" in Rizzolatti and Sinigaglia, *Mirrors in the Brain*, 115–38.

47. For a summary of these well-known effects, which I analyze further below, see Bastiaansen et al., "Evidence for Mirror Systems in Emotions."

48. Iacoboni, *Mirroring People*, 216. See B. Calvo-Merino, Daniel E. Glaser, et al., "Action Observation and Acquired Motor Skills: An fMRI Study with Expert Dancers," *Cerebral Cortex* 15 (2005): 1243–49. Also see Catherine Stevens and Shirley McKechnie,

"Thinking in Action: Thought Made Visible in Contemporary Dance," *Cognitive Process* 6 (2005): 243–52.

49. See B. Calvo-Merino, Julie Grèzes, et al., "Seeing or Doing? Influence of Visual and Motor Familiarity in Action Observation," *Current Biology* 16 (2006): 1905–10.

50. Rizzolatti and Sinigaglia, *Mirrors in the Brain*, 34. Rizzolatti attributes the notion of affordance to James J. Gibson, *The Ecological Approach to Visual Perception* (Boston: Houghton Mifflin, 1979).

51. David Freedberg and Vittorio Gallese, "Motion, Emotion, and Empathy in Esthetic Experience," *Trends in Cognitive Sciences* 11.5 (March 2007): 200.

52. Zahavi, *Subjectivity and Selfhood*, 163. For Heidegger's analysis of "equipment" and the "ready-to-hand" (*Zuhandenheit*), see *Being and Time*, 102–7.

53. Martin Heidegger, "The Origin of the Work of Art" (1935–36), in *Poetry, Language, Thought*, trans. Albert Hofstadter (New York: Harper & Row, 1971), 15–87.

54. Cinzia Di Dio and Vittorio Gallese, "Neuroaesthetics: A Review," *Current Opinion in Neurobiology* 19 (2009): 683. Di Dio and Gallese discuss Cinzia Di Dio et al., "The Golden Beauty: Brain Response to Classical and Renaissance Sculptures," *PloS ONE* 11 (2007): e1201.

55. For example, according to Rizzolatti and Sinigaglia, "we know that Broca's area, one of the classic areas of language, also possesses motor properties that are not exclusively verbal, becoming active during oro-facial, brachio-manual, and oro-laryngeal movements, and that its organization is similar to that found in the homologous area in monkeys, i.e. F5. Moreover, Broca's area, just as F5, is a part of a mirror neuron system whose primary function is to link action understanding with action production in both humans and monkeys." *Mirrors in the Brain*, 159.

56. See Iacoboni, *Mirroring People*, 90. Hickok contests the connection between language and motor responses based on the absence of evidence that lesions to Broca's area impair action understanding (see "Eight Problems," 1237), although he concedes that this is not conclusive, because other areas may compensate for this deficiency. Iacoboni's TMS evidence would seem to answer this objection.

57. See Olaf Hauk and Friedman Pulvermüller, "Neurophysiological Distinction of Action Words in the Fronto-Central Cortex," *Human Brain Mapping* 21.3 (2004): 191–201; and Véronique Boulenger et al., "Cross-Talk Between Language Processes and Overt Motor Behavior in the First 200 msec of Processing," *Journal of Cognitive Neuroscience* 18.10 (2006): 1607–15. Also see David S. Miall's discussion of these and other corroborating experiments in "Neuroaesthetics of Literary Reading," in *Neuroaesthetics*, ed. Martin Skov and Oshin Vartanian (Amityville, NY: Baywood, 2009), 242.

58. See, e.g., Hickok, "Eight Problems," 1229–30, 1238–39. On the relation between the structuralist perspective on language as a formal system of rules and the phenomenological view of meaning as an embodied event, see Paul Ricoeur, "Structure, Word, Event," trans. Robert Sweeney, in *The Conflict of Interpretations*, ed. Don Ihde (Evanston, IL: Northwestern UP, 1974), 79–96.

59. Roman Ingarden, *The Literary Work of Art* (1931), trans. George G. Grabowicz (Evanston, IL: Northwestern UP, 1973), 125–26.

60. Paul Ricoeur, *Time and Narrative*, trans. Kathleen McLaughlin and David Pellauer (Chicago: U of Chicago P, 1984), 1:142; see also 1:52–87.

61. Kenneth Burke, *Language as Symbolic Action* (Berkeley: U of California P, 1966).

62. Aristotle, *Poetics*, trans. Hippocrates G. Apostle et al. (Grinnell, IA: Peripatetic Press, 1990), 6, 9.

63. Cristina Becchio and Cesare Bertone, "Beyond Cartesian Subjectivism: Neural Correlates of Shared Intentionality," *Journal of Consciousness Studies* 12.7 (2005): 20.

64. Gallese, "'Shared Manifold' Hypothesis," 38: "It can be hypothesized that echopractic behavior represents a 'release' of a covert action simulation present also in normal subjects, but normally inhibited in its expression by the cortical areas that in these patients are functionally defective."

65. Keysers reports that recent experiments have also discovered a new category of cells, *anti-mirror neurons*, which "increased their firing rate when the patient was executing a particular action, but *decreased* their firing rate below baseline when the patient observed someone else perform this action. . . . Anti-mirror neurons could disambiguate our own actions from those of others." "Social Neuroscience," R354, emphasis in original.

66. See Rizzolatti and Sinigaglia, *Mirrors in the Brain*, 151–52. Rizzolatti suggests that newborns may exhibit compulsive and precocious imitative activity because "they already possess a mirror neuron system, albeit rather rudimentary, and . . . their control mechanisms are still weak" due to the still "modest functionality of the frontal lobe" (152).

67. Susanne Küchler similarly conceptualizes "skin" as "a psychosomatic membrane that forms both a boundary and a point of contact between the inner and outer worlds" in her fascinating essay "The Extended Mind: An Anthropological Perspective on Mind, Agency, and 'Smart' Materials," in *A Field Guide to a New Meta-Field: Bridging the Humanities-Neurosciences Divide*, ed. Barbara Maria Stafford (Chicago: U of Chicago P, 2011), 86.

68. On the literary implications of empathy, see Suzanne Keen, *Empathy and the Novel* (New York: Oxford UP, 2007). Keen mentions mirror-neuron research, however, only to dismiss it: "This newly enabled capacity to study empathy at the cellular level encourages speculation about human empathy's positive consequences. These speculations are not new, as any student of eighteenth-century moral sentimentalism will affirm" (viii). Perhaps not, but my analysis of the neurobiological research will, I think, give additional reasons for the skepticism Keen wisely voices about the ability of literary reading experiences to inculcate moral compassion, and as she shows, that is a question that eighteenth-century psychology did not exhaust. Miall offers brief but perceptive observations about the possible relevance of mirror-neuron research to empathic reading experiences in "Neuroaesthetics of Literary Reading," 240–44. As is unfortunately typical of recent cognitive literary studies, neither Keen nor Miall shows any awareness of the phenomenological tradition's analysis of these questions.

69. Martin L. Hoffman, *Empathy and Moral Development: Implications for Caring and Justice* (Cambridge: Cambridge UP, 2000), 3.

70. Jennifer H. Pfeifer and Mirella Dapretto, "'Mirror, Mirror, in My Mind': Empathy, Interpersonal Competence, and the Mirror Neuron System," in *The Social Neuroscience of Empathy*, ed. Jean Decety and William Ickes (Cambridge, MA: MIT P, 2009), 184. On the ambiguities of empathy, see C. Daniel Baston, "These Things Called Empathy: Eight Related but Distinct Phenomena," in Decety and Ickes, *Social Neuroscience of Empathy*, 3–15; and Tania Singer and Susanne Leiberg, "Sharing the Emotions of Others: The Neu-

ral Bases of Empathy," in *The Cognitive Neurosciences*, ed. Michael S. Gazzaniga, 4th ed. (Cambridge, MA: MIT P, 2009), 973–86.

71. See Theodor Lipps, "Einfühlung, innere Nachahmung und Organempfindungen," in *Archiv für die Gesamte Psychologie*, vol. 1, pt. 2 (Leipzig: W. Engelmann, 1903), 185–204. Also see Iacoboni's account in *Mirroring People*, 108–9.

72. Hein and Singer, "I Feel How You Feel," 154.

73. Singer and Leiberg, "Sharing the Emotions of Others," 974.

74. See, e.g., Jean Decety and Claus Lamm, "Empathy Versus Personal Distress: Recent Evidence from Social Neuroscience," in Decety and Ickes, *Social Neuroscience of Empathy*, 199–213, from which the quotations in the text are taken.

75. Baston, "These Things Called Empathy," 10.

76. Decety and Lamm, "Empathy Versus Personal Distress," 208.

77. See Keen's critique of Martha Nussbaum and other traditional advocates for the ability of literature to promote moral conduct in *Empathy and the Novel*, esp. 37–64.

78. See W. D. Hutchison et al., "Pain-Related Neurons in the Human Cingulate Cortex," *Nature Neuroscience* 2 (1999): 403–5. Also see Rizzolatti and Sinigaglia's discussion in *Mirrors in the Brain*, 187.

79. See Tania Singer et al., "Empathy for Pain Involves the Affective but Not Sensory Components of Pain," *Science* 303 (2004): 1157–62. Also see Rizzolatti and Sinigaglia, *Mirrors in the Brain*, 187.

80. Yawei Cheng, "Love Hurts: An fMRI Study," *NeuroImage* 51 (2010): 927.

81. Pfeifer and Dapretto, "Mirror, Mirror, in My Mind," 191.

82. Hein and Singer, "I Feel How You Feel," 156. For the report of the experiment, see Yawei Cheng et al., "Expertise Modulates the Perception of Pain in Others," *Current Biology* 17 (2007): 1708–13.

83. Bruno Wicker et al., "Both of Us Disgusted in My Insula: The Common Neural Basis of Seeing and Feeling Disgust," *Neuron* 40 (2003): 655. For a provocative literary-historical analysis of this fundamental emotion, see Winfried Menninghaus, *Disgust: The Theory and History of a Strong Sensation*, trans. Howard Eiland and Joel Golb (Albany: SUNY P, 2003).

84. Antonio Damasio, *Looking for Spinoza: Joy, Sorrow, and the Feeling Brain* (New York: Harcourt, 2003), 115–16.

85. See Antonio Damasio, *Descartes' Error: Emotion, Reason, and the Human Brain* (New York: Putnam, 1994), 155–64.

86. Bastiaansen et al., "Evidence for Mirror Systems in Emotions," 2391. He warns that "there is little evidence for a consistent mapping of particular emotions onto particular brain regions. Instead, different networks seem to be involved dependent on the process by which the emotion is accessed" (2393).

87. See James Elkins, *Pictures and Tears: A History of People Who Have Cried in Front of Paintings* (New York: Routledge, 2001).

88. See Friedrich Nietzsche, *The Birth of Tragedy Out of the Spirit of Music* (1872), trans. Shaun Whiteside (New York: Penguin, 1994). Also see Leo Tolstoy's depiction of the visceral, embodied effects of music in his novella *The Kreutzer Sonata* (1889).

89. See Elaine Scarry's eloquent reflections about the differences in "immediate sen-

sory content" between painting, music, and literature in *Dreaming by the Book* (New York: Farrar Straus Giroux, 1999).

90. Andrew N. Meltzoff and Jean Decety, "What Imitation Tells Us About Social Cognition: A Rapprochement Between Developmental Psychology and Cognitive Neuroscience," *Philosophical Transactions of the Royal Society London B* 358 (2003): 491. The foundational work on which this paper reports includes the following by Meltzoff and his collaborator M. K. Moore: "Imitation of Facial and Manual Gestures by Human Neonates," *Science* 198 (1977): 75–78; "Newborn Infants Imitate Adult Facial Gestures," *Child Development* 54 (1983): 702–9; and "Imitation in Newborn Infants: Exploring the Range of Gestures Imitated and the Underlying Mechanisms," *Developmental Psychology* 25 (1989): 954–62.

91. Gallese, "Manifold Nature of Interpersonal Relations," 518.

92. On play and the "as if," see Roger Caillois, *Man, Play, and Games* (1958; Urbana: U of Illinois P, 2001), esp. 3–35; and Wolfgang Iser, *The Fictive and the Imaginary: Charting Literary Anthropology* (Baltimore: Johns Hopkins UP, 1993), esp. 247–80. Also see my essay "The Politics of Play: The Social Implications of Iser's Aesthetic Theory," *New Literary History* 31.1 (Winter 2000): 205–17.

93. Sir Philip Sidney, "An Apology for Poetry" (1583), in Adams, *Critical Theory Since Plato*, 168.

94. Ramachandran, *Tell-Tale Brain*, 137. In addition to Simon Baron-Cohen's classic study of ToM deficiencies associated with autism in *Mindblindness* (see n. 5 above), see Singer and Leiberg's survey of the current state of research in "Sharing the Emotions of Others," 979. Also see Uta Frith, "Mind Blindness and the Brain in Autism," *Neuron* 32 (2001): 969–79; and G. Silani et al., "Levels of Emotional Awareness and Autism: An fMRI Study," *Social Neuroscience* 3 (2008): 97–112.

95. Marco Iacoboni, "Understanding Others: Imitation, Language, and Empathy," in *Perspectives on Imitation*, vol. 1, *From Neuroscience to Social Science—Mechanisms of Imitation and Imitation in Animals*, ed. Susan Hurley and Nick Chater (Cambridge, MA: MIT P, 2005), 108, 109.

96. On the opposition between viewing language as structure and viewing it as intersubjective speech, also see Ricoeur, "Structure, Word, Event" (see n. 58 above).

97. Gadamer consequently argues that "not just occasionally, but always, the meaning of a text goes beyond its author" because "we understand in a different way, if we understand at all." *Truth and Method* (1960), trans. Joel Weinsheimer and Donald G. Marshall, 2nd ed. (New York: Continuum, 1993), 296, 297. On the arguments surrounding authorial intention, see my *Conflicting Readings: Variety and Validity in Interpretation* (Chapel Hill: U of North Carolina P, 1990), esp. 1–19.

98. Michael Tomasello et al., "Understanding and Sharing Intentions: The Origins of Cultural Cognition," *Behavioral and Brain Sciences* 28 (2005): 676.

99. On the social implications of the experience of reciprocity in reading, see my *Play and the Politics of Reading: The Social Uses of Modernist Form* (Ithaca, NY: Cornell UP, 2005), esp. 2–21.

100. Richard J. Gerrig and Philip G. Zimbardo, *Psychology and Life*, 17th ed. (Boston: Pearson, 2005), 200, 199. They cite the classic 1963 study by Albert Bandura et al., "Imitation of Film-Mediated Aggressive Models," *Journal of Abnormal and Social Psychology* 66.1

(1963): 3–11. Also see George Comstock and Erica Scharrer, *Television: What's On, Who's Watching, and What It Means* (San Diego: Academic, 1999).

101. Craig A. Anderson et al., "Violent Video Game Effects on Aggression, Empathy, and Prosocial Behavior: A Meta-Analysis," *Psychological Bulletin* 136.2 (2010): 151; L. Rowell Huesmann, "Nailing the Coffin Shut on Doubts that Violent Video Games Stimulate Aggression: Comment on Anderson et al. (2010)," *Psychological Bulletin* 136.2 (2010): 179.

102. Christopher J. Ferguson and John Kilburn, "Much Ado About Nothing: The Misestimation and Overinterpretation of Violent Video Game Effects in Eastern and Western Nations: Comment on Anderson et al. (2010)," *Psychological Bulletin* 136.2 (2010): 176.

103. Daphne Bavelier et al., "Exercising Your Brain: Training-Related Brain Plasticity," in Gazzaniga, *Cognitive Neurosciences*, 155.

104. See René Girard, *Deceit, Desire, and the Novel: Self and Other in Literary Structure*, trans. Yvonne Freccero (Baltimore: Johns Hopkins P, 1965); and Girard, *Violence and the Sacred*, trans. Patrick Gregory (Baltimore: Johns Hopkins UP, 1977). The research on mirror neurons and infant imitation has brought renewed interest to Girard's work on mimesis and violence. See Scott R. Garrels, ed., *Mimesis and Science: Empirical Research on Imitation and the Mimetic Theory of Culture and Religion* (East Lansing: Michigan State UP, 2011).

105. See my *Phenomenology of Henry James* (see n. 20 above), esp. 138–40, for an analysis of Sartre's claim that conflict is the original meaning of the relation between self and other, as opposed to Heidegger's analysis of care (*Sorge*) as the fundamental structure of "being with" (*Mitsein*).

EPILOGUE

1. From 1994 to 1996 I was an associate dean in the College of Arts and Sciences at the University of Oregon. I then served as dean of the College of Arts and Sciences at the State University of New York at Stony Brook from 1996 to 2000, after which I was dean of the College at Brown from 2001 to 2006.

2. This is also known as the problem of "emergence"—how physical, electrochemical processes can give rise to consciousness and experience. For a recent provocative attempt to solve this problem, see Terrence W. Deacon, *Incomplete Nature: How Mind Emerged from Matter* (New York: Norton, 2012).

3. John R. Searle, "Consciousness," *Annual Review of Neuroscience* 23 (2000): 567.

4. Thomas Nagel, "What Is It Like to Be a Bat?," *Philosophical Review* 83 (1974): 442, emphasis in original.

5. Francis Crick, *The Astonishing Hypothesis: The Scientific Search for the Soul* (New York: Simon & Schuster, 1994), xii.

6. See Martin Heidegger, *Being and Time* (1927), trans. John Macquarrie and Edward Robinson (New York: Harper & Row, 1962), 67–68. On the ethical and aesthetic implications of *Jemeinigkeit* for reading, see J. Hillis Miller's interesting essay "Should We Read 'Heart of Darkness'?" in *Conrad in Africa: New Essays on "Heart of Darkness,"* ed. Attie de Lange and Gail Fincham with Wiesław Krajka (New York: Columbia UP, 2002), 21–39. "No one can do your reading for you," Miller argues. "Each must read again in his or her turn and bear witness to that reading in his or her turn" (21). The aesthetic experience

is "own-most," then, as are its ethical implications, that is, how the reader responds to the call of this experience to take up certain responsibilities (how we "bear witness" to what we have experienced). This is one of the reasons why the moral value(s) of a literary work can be ambiguous, even if it is unambiguously the case that literature has ethical implications.

7. See the discussion in chapter 2 of Changizi's research on cross-alphabetic graphic patterns and the "kiki"-"bouba" experiments.

8. See the discussions in chapter 2 of neuronal recycling and reading, in chapter 3 of the neuroscientific foundations of the hermeneutic circle, and in chapter 4 of how the evolutionary time scales of reading challenge claims about the historical relativity of reading practices.

9. Edward O. Wilson, *Consilience: The Unity of Knowledge* (New York: Vintage, 1998), 60.

10. Wilson's own, unfortunately simplistic statements about art provide compelling evidence of science's need for the explanatory powers of the humanities, for example: "Art is the means by which people of similar cognition reach out to others in order to transmit feeling." *Consilience*, 128. Art involves much more than the transmission of feeling, of course, and the reference to people of similar cognition slides over the interesting and important questions raised by the different ways that critics and communities have interpreted the same text. No undergraduate who has taken my course "The History of Criticism from Plato to Postmodernism" would venture such a sweeping, questionable generalization. Scientists need to draw on the expertise of humanists with the same cross-disciplinary caution and humility they would show in encroaching on another scientific field.

11. See my essay "Form and History: Reading as an Aesthetic Experience and Historical Act," *Modern Language Quarterly* 69 (June 2008): 195–219.

Index

abstract art, 85
action potential, 103–4, 106, 113, 199–200n34
action understanding, 138–39, 141–42, 149–51, 153–58, 166–67
acupuncture, 160
addiction, 50
Adorno, Theodor W., 45, 49
aesthetic emotions, 11, 17, 43, 125, 161, 164–65
aesthetic experience, x–xi, xiv, 1–2, 9, 174; doubling in, 102, 151, 156, 159, 161, 169–72; harmony and dissonance in, 21–22, 26–27, 84; heterogeneity of, 11–13, 16–18, 47, 52, 181; neural correlates of, 5–7, 42–43, 51, 177–78; surprise in, 86–87, 171
aesthetics, 1, 6–7, 10–11, 16–17, 20–21, 37
affect theory, 185n19
affordance, 149–50
agency, 24, 150, 153
aggression, 173–74
alphabets, 28, 30–32, 46, 58
ambiguous handwriting, 79
ambiguity, xi, 3, 195n24; and interpretive conflict, 21, 42, 52, 55–58, 81; of multi-stable images, 67–72, 74–75, 89–90
Anderson, Craig, 173
anticipation and retrospection, 54, 57, 97, 101, 103
anticipatory understanding, 56, 81, 123
Aristotle, 1, 11, 125, 154–55, 178
art: vs. non-art, 3, 14, 16–17, 22, 47, 71; works of, 150–51
artists as neuroscientists, 90
"as if," the, 17, 146, 155, 161–65, 168–69, 172
as-if body loop, xi, 11, 125–26, 163–64
"as" relations, 167, 169, 173
astonishing hypothesis, 177
attention, 68, 70–71, 76, 89, 107, 110, 195n26

attunement (*Stimmung*), 123–24, 202n69
auditory system, 14–15, 32, 39, 44–45, 81, 106
Augustine, 91, 197n1
Austen, Jane, 46; *Pride and Prejudice*, 198n16
autism, 170, 210n94
avant-garde, 5, 45–46, 49

Baars, Bernard, 92, 99, 104, 108–9
Balzac, Honoré de, 46, 143
Bandura, Albert, 210n100
baseball, 4, 6, 180
Bastiaansen, J. A. C. J., 164
Baston, Daniel, 159
bats, 99
Bavelier, Daphne, 174
Bear, Mark, 104–5
beauty, 37, 42–43, 49, 51; facial, 38–39
Becchio, Cristina, 157
Beckett, Samuel, 46, 143
Beethoven, Ludwig van, 15, 49; Fifth Symphony, 96–97
Bell, Clive, 17
Bertone, Cesare, 157
Biederman, Irving, 51
billiard-ball causality, 79, 82–83, 176
binding problem, 60, 100, 105–9, 121, 197n45
biological naturalism, 177, 180
Blake, William, 186n29
Bloom, Paul, 125
the body. *See* embodiment
boredom vs. overstrain, 86
"bouba"-"kiki" experiment, 32
Boston Red Sox, 180
Boston Symphony Orchestra, 128
Boyd, Brian, 34, 111, 147, 193n58
Boyle's air pump, 8
bracketing, 20

brain: asynchronous processing, 60, 66–69, 98–101, 127, 199n22; "bushy," 2–3, 33, 41–42, 54; chemistry, 46–47; as cocktail party, 107, 111, 127–28; decentered, x, 2–3, 33, 42, 48, 52, 100, 102, 108, 121, 126–30; instability, 67, 75, 101–2, 107, 120, 129; localization, 12, 17, 26, 36, 38, 40–41, 43, 60, 115, 121, 129, 147, 164, 191n35, 194n22, 209n86; mapping in, 106, 121, 125, 166–68, 170; reward system, 37, 42, 178; rhythms, 8, 100, 105–9, 111; right vs. left hemispheres, 72–73, 122, 202n67; social, 19, 131–36, 144, 171–74; as symphony, 127–28; in a vat, 133, 204n13

brain, anatomy of: amygdala, 50, 106, 125; anterior cingulate cortex (ACC), 146–47, 160; Broca's area, 96, 151–52, 207n55; hippocampus, 39–40, 50, 72, 115; insula, 146–47, 160, 162–63; motor cortex, 121–22, 154; prefrontal cortex, 50; stem, 121–22, 164; striatum, 50, 115; thalamus, 121, 164; ventral tegmental area (VTA), 50; visual cortex, 40, 51, 59–61, 63, 69, 70, 106; Wernicke's area, 96

brain-body divide, 163, 165
Brecht, Bertolt, 13
bricolage, 35
Brontë, Charlotte, *Jane Eyre*, 142–43, 147
Brook, Andrew, 18
Brooks, Cleanth, 15
Buddha, 1
buffering, 96
Burke, Kenneth, 7, 155
Buzsáki, György, 73

Campion, Jane, *Portrait of a Lady* (film), 85
canonical neurons, 149, 150–51, 153
Cartesian model, 77, 129, 131
catharsis, 11, 125
center/surround structure, 65–66
central controller, 41, 107, 121, 127–28, 191n35
Changeux, Jean-Pierre, 13, 24
Changizi, Mark, 30–32
Chatterjee, Anjan, 38
Cheng, Yawei, 160
Chomksy, Noam, 153, 188n1, 196n38
Churchland, Patricia, 131, 139, 141, 172, 192n48
classical aesthetics, 13, 22, 49–50, 110
classical conditioning, 114
climate change, 8
closure, 52

coding, 104–5, 108
cognition, 77, 120
cognitive literary studies, xii, 18, 185n23
cognitive perspective taking, 146–47
color perception, 12, 35, 59, 63–64, 70, 74, 195n30
computer model, 20, 27, 41, 54, 78–82, 191n35
conceptual blending, 87–88
concretization, 21, 98
cones, 61–64
conflict of interpretations, x–xi, xiv; and ambiguous texts, 55, 57–58; basis in brain, 21, 67, 71, 73–75, 77–78, 116, 201n56; between ways of reading, 2, 5, 7, 9, 85–86, 210n97
Connors, Barry, 104–5
Conrad, Joseph, 103, 143; *Lord Jim*, 103, 146
consciousness, 5, 7, 128–29
consilience, 179
consistency building: and breaking, 86–88; in perception, 51, 62, 75–76, 89; in reading, x, 2, 21, 54, 84–86, 117, 148, 153–54, 196n40; temporality of, 92, 97, 103, 106, 116
consonance, 14, 44–45, 49
constancy, 39, 84, 92; color, 64, 89; consistency building, 68–69, 73–78, 86–88, 107, 113; and flexibility, ix, 3, 9, 57, 65, 73–74, 78, 113; and novelty, 65, 86, 92, 107, 113
context, 55, 79
contextual criticism, 10, 21
conventions, 26, 32, 44, 46, 55, 113
correlation, 178
Cratylism, 32, 189n9
Crick, Francis, 177
cultural differences, 35, 37, 39, 45, 121, 165, 178
cultural objects, 150

Damasio, Antonio, xi, 76, 101, 106, 120–22, 125–28, 163–64, 203n77
Dames, Nicholas, 119
dance, 100, 149
Dapretto, Mirella, 158
Darnton, Robert, 116–18
Darwin, Charles, 34, 53, 190n15; natural selection, 27
David, Larry, 5
Decety, Jean, 159
deconstruction, 13, 24
deep orthographies, 34
defamiliarization, x, 13, 22, 48, 50, 111, 113, 145

Dehaene, Stanislas, x, 13, 19, 27–29, 33, 40–41, 78–79, 81, 117–19
de Man, Paul, 15
depolarization and repolarization, 103
Derrida, Jacques, 188n54
Devlin, Joseph, 33
Dickens, Charles, 46; *The Old Curiosity Shop*, 146–47
Di Dio, Cinzia, 151
Dinstein, Ilan, 148
direct-matching hypothesis, 138
disgust, 149, 161–64
disinterestedness, 17, 51–52
Disney, Walt, *Fantasia*, 49
disorientation and reorientation, 87–88
dissonance: aesthetics of, xiii–xiv, 13–16, 22–23, 26–27; in ambiguity, 69, 75–76; and defamiliarization, 48–51, 111, 119, 181; in metaphor, 87–89; in neural flexibility, ix, 67; and surprise, 43–44, 124
distortion, 13
Donald, Merlin, 131
dopamine, 50–51, 53
Dostoevsky, Fyodor, *Crime and Punishment*, 135–36
doubling, 102, 135–36, 138, 146, 150, 155–61, 165, 168–74
Dreyfus, Hubert, 81, 191n35, 193n4
duck-rabbit figure, 55–58, 67–68, 107–8
dyslexia, 36

echopraxia, 157, 208n64
electrical signatures, 103–5
electroencephalogram (EEG), 105, 108, 124, 148
eliminativism, 19
Eliot, George, 46, 143
Eliot, T. S. (Thomas Stearns): "The Love Song of J. Alfred Prufrock," 88; "objective correlative," 85
Elkins, James, 164
embodiment, 77, 120–22, 124–26, 138, 150–54, 161–65, 170
emergence, 211n2
emotion: and the body, 120–22, 125–26, 144–49, 163–65, 209n86; and cognition, 76–77; in empathy, 158–59, 161–65, 170; and the future, 122–24; in reading, 144–48
empathy, 18, 146, 158–62, 165, 170
emplotment, 154

enactment imagination, 144
engram, 116
entertainment art, 23
epilepsy patients, 148
epileptic seizures, 108–10
Ethington, Philip J., 201n58
evolution, 8, 26, 28, 30, 32, 63, 193n58; adaptations, 34–36, 53; cultural, 172; and emotions, 121; and reading, 117–18, 131, 179, 188n1
expectations: in interpretation, 11, 55–57, 77, 124; in reading, 3, 79, 81, 84–87, 97; violations of, 23, 43–44, 94, 101
expertise, 112, 115–16
explanatory gap, xi, xiii, 7–8, 10, 12, 19–21, 178–79

face recognition, 12, 37–38
Faulkner, William, 143; *Sound and the Fury*, 103
fauvist painting, 89
fear, 149, 163–64
Ferguson, Christopher, 173
film, 5, 165, 172
first-person experience, 6, 20, 126
Fischer, Steven Roger, 27
Flaubert, Gustave, 14; *Madame Bovary*, 15
flexibility, 14, 48, 50, 64, 67–68, 84, 86–88, 120. *See also* constancy
focalization, 146
Fodor, Jerry A., 191n35
Ford, Ford Madox, *The Good Soldier*, 103
foregrounding, 16, 124
form, 151, 172, 181–82; literary, 145
formalism, 181–82
fovea, 59
Freedberg, David, 149, 151, 164
Firth, Chris and Uta, 146–47
functional magnetic resonance imaging (fMRI), 4–5, 16, 22, 29, 105, 131, 148, 151, 184n6, 206n44
futurity, 101, 122–24, 129, 199n25. *See also* protentional horizon

Gadamer, Hans Georg, 21, 51, 87, 112, 210n97
Gage, Nicole, 92, 99, 104, 108–9
Gage, Phineas, 76
Gallagher, Shaun, 91, 93, 115, 123–24, 126
Gallese, Vittorio, 133, 149, 151, 157, 164, 166–68, 205n25, 208n64
ganglion cells, 61–62, 194n17
gap filling, x, 2, 69, 84–85, 116, 153

Gastaut, Henri, 148
Gazzola, Valeria, 132
Genette, Gérard, 102
genre, 54–55
Gerrig, Richard, 86, 173, 196n40
gestalt shifts, 55, 67, 70–71, 108
gesture, 153
Girard, René, 174, 211n104
Goldberg, Elkhonon, 72, 89
Goldman, Alvin, 124, 144
Gombrich, E. H., 55, 67–68
Goodman, Nelson, 87
gossip, 144–45
Goswami, Usha, 33
Gould, Stephen Jay, 35
grapheme-phoneme translation, 27, 32, 34–35, 41–42, 67, 80–81

habit, 14, 48–50, 73, 86–87, 113, 145; formation, 111–12, 115, 118–20
habituation, 112–15, 120
Handel, George Frideric, 114
Hardcastle, Valerie Gray, 40
hard problem, xi, 5, 7, 11, 24, 176
harmony: aesthetics of, xiii–xiv, 13–16, 22–23, 26–27, 124, 178, 181; in music, 14, 44–46, 96, 110; in neural syntheses, ix, 14, 43–51, 67, 84, 86, 92, 110–11; and sensitization, 114–15
Hasson, Uri, 5
Haydn, Joseph, 114–15
Hayles, N. Katherine, 185n19
hearing, 26, 58, 166. *See also* auditory system
Hebb's Law, 39, 40, 73, 86, 116–18, 190n27
Heidegger, Martin, 101, 202n69; on gossip (*Gerede*), 144; on interpretation, 21, 56, 123; "my-own-ness" (*Jemeinigkeit*), 177, 211–12n6; "The Origin of the Work of Art," 150; on tools (*Zuhandenheit*), 150
Hein, Grit, 146, 159, 161
heresy of paraphrase, 176
hermeneutic circle, x, 3–4, 11, 21, 54–59, 65, 82, 84, 90, 92, 141, 143; anomalies, 79; familiar and unfamiliar, 72–73, 119, 142, 167; parts and wholes, 63–64, 77–78, 80–82, 86–88, 97, 117, 176; selection and combination, 62, 64; or spiral, 56–57, 92; vicious circularity, 57, 64, 76
Hickok, Gregory, 139, 141–42
Hirsch, E. D., Jr., 77, 196n32

historical relativity, 2, 8, 14–17, 32, 37, 116–18, 165
historicism, 181–82
Hitchcock, Alfred, 5
Hoffman, Martin L., 158
Hogan, Patrick Colm, 16, 186n32
Holland, Norman N., 3, 183n5, 185n23
homeostasis, 101
homonyms, 42
homunculus, 23, 25, 42, 60, 74, 77, 126
honey bees, 106
Horatian maxim, 47, 51, 118, 169
horizon, 95, 97, 107, 127. *See also* protentional horizon; retentional horizon
Hruby, George, 33
Huesmann, Rowell, 173
humanities, the, 21, 120, 172; interpretive conflict in, 74, 78, 116; and the sciences, 1–2, 8–11, 174–77, 180–82, 212n10
Husserl, Edmund, xii, 20, 126; on intersubjectivity, 133–35; on temporality, 20, 91, 93–95, 107
Hutchison, William, 160
Hyman, John, 6, 183n2
hypersynchrony, 108–9, 111

Iacoboni, Marco, 131–33, 149, 152–53, 157, 160, 170, 172
identification, 146, 156, 158–59, 161, 165
illusion building, 84, 143, 145, 148, 153
imitation, 154–56, 165, 173–74; embodied, 152–56; infant, 19, 165–69, 171, 208n66; and violence, 172–73
indeterminacies, 85–86, 153; and the future, 123–24
inertia, 119
infants: face recognition, 38; consonance, 44. *See also* imitation, infant
Ingarden, Roman, 16, 20–22, 98, 110, 153
inherited vs. acquired traits, 35–38, 40
instrumental games, 52
intentionality, 21, 95, 102; derived, 153–54, 158, 171; shared ("we"), 171–72
intentions, 139, 141, 205n28; authorial, 210n97
interaural time delay, 99
interdisciplinary study, 1, 10–11, 180–81
interpretation, x, xiv, 3, 20–21, 54, 92; "as" structure, 167, 169; forestructure (*Vorstruktur*), 56, 123. *See also* hermeneutic circle

intersubjectivity, xi, 21, 131–35, 153, 156, 205n25; primary, 133–34, 136, 138–39, 142, 145, 150, 168
invariant object recognition, 28, 30, 32, 39, 42, 58
irony, 15
Iser, Wolfgang, 21; on consistency building, 21, 119, 143; on dissonance, 22, 87, 89; on gaps, 7, 84–85; on play, 51–52; on self-other relation, 135, 161, 171; on temporality, 92, 97

Jacob, Pierre, 205n28
Jakab, Zoltán, 59
Jakobson, Roman, 13, 16, 52
James, Henry, xii, 60, 85, 135, 145; *The Ambassadors*, 147; *The Golden Bowl*, 146; *Portrait of a Lady*, 165; *The Sacred Fount*, 195n24; *Turn of the Screw*, 3, 55, 58; *The Wings of the Dove*, 147
James, William, xii, 48, 65, 94, 127, 129, 199n27; *Principles of Psychology*, xii, 91
Jauss, Hans Robert, 21, 23, 49, 87, 111, 118
"Jennifer Aniston neuron," 12, 35, 38, 40
Joyce, James, 112, 143; *Finnegans Wake*, 86; *A Portrait of the Artist as a Young Man*, 146; *Ulysses*, 16, 46, 86, 147
judgment, 76
juggling, 39

Kafka, Franz, *The Trial*, 198n16
Kandel, Eric, 113, 201n53
Kanizsa triangle, 68–69, 70
Kant, Immanuel, x, 1, 17, 51
Keen, Suzanne, 160, 201n61, 208n68
Keller, David, 12
Kermode, Frank, 102
Keysers, Christian, 132, 148, 206n44
Kidman, Nicole, 85
Kierkegaard, Søren, 102, 199n27
Kilburn, John, 173
Konstanz School, 21–23, 183n4
Kroeber, Alfred, 83

landscapes, 37
Langlois, Judith, 38
language, 96, 153, 156, 158; acquisition, 27, 39, 168, 170, 188n1; as action, 151, 154–56; and music, 96
Latour, Bruno, 8, 184–85n18
Lawrence, D. H., 144

learning, 45–47, 72–73, 89; and habit formation, 113–16, 118–19; and imitation, 168–72; observational, 173; to read, 23, 25–26, 33–41
Lehrer, Jonah, 45, 49, 185n22, 197n46
lesions, 36, 41, 60, 76, 96, 122, 152
letterbox area. *See* visual word form area
Levine, James, 128
Lévi-Strauss, Claude, 35
Levitin, Daniel, 44, 96, 99, 105
Linden, David J., 50
linear parsing, 196n38
linguistic recursivity, 82–84
Lipps, Theodor, 158–59
literacy, 81, 118, 131
literariness, 111
literary theory, 6–7, 82, 179
literary universals, 16, 186n32
literature: effects of, 118–20; heterogeneity of, 1, 14, 16–17, 120, 181
Livingstone, Margaret, 58–59, 61–63, 65–66, 194n17
London taxi drivers, 39
long-distance dependencies, 82, 84
love, 160; romantic and maternal, 37–38, 47
Lukács, Georg, 13

Machiavelli, Niccolò, 174
Malkovich, John, 85
Mandik, Peter, 19
Martindale, Colin, 107
Marxist aesthetics, 13
Matthews, Gary, 194n17, 200n35
McGinn, Colin, 5
McIlwain, Jim, 199n34
meaning vs. significance, 77, 196n32
melody, 95–96
Meltzoff, Andrew N., 165–70
memory: and interpretation, 51, 72; kinds of, 94, 115; and learning, 113–16; and reading, 32, 97–98, 196n40; temporality of, 93–94, 100
Merleau-Ponty, Maurice, xii, 202n69; on paradox of alter ego, 133–34, 150, 153; on temporality, 91, 102, 127, 129
metaphor, 23–24; dead, 88; interaction theory of, 87–89
metapicture, 68
Miall, David, 16, 111, 124, 185n23, 186n34, 208n68
Miller, J. Hillis, 211–12n6

mimetic desire, 174
mind-brain divide, xiii, 18–19, 126
mind reading, 132–33. *See also* theory of mind
mirror metaphor of mind, 59–60
mirror neurons (MNs), 8, 19, 132–33, 136–44,
    163; anti-mirror-neurons, 208n65; con-
    troversies about, 139, 141–42, 205n28; in
    humans, 148–49, 152, 206n45; in imitation,
    165–68, 172–73; and intersubjectivity, xi,
    19, 133, 151–53, 156–57, 160, 170; kinds of,
    137–38, 148. *See also* canonical neurons
mirror symmetry, 36–37
Mitchell, W. J. T., 68, 89
modular mind, 191n35
Mona Lisa, 49
monism, 17
monkeys, 38, 160; macaque, 44–45, 51, 132–33,
    137–40, 148, 151–52, 157, 207n55
Morais, José, 81
morality, 118–20, 159–60, 172, 174, 192n48
motion, 35; blindness (*akinetopsia*), 92–93
motor resonance. *See* action understanding
motor system, 4, 39, 154–56
multi-stable images, 68, 71, 74, 78, 89
music, 26, 108, 164–65; atonal, 45–46, 115;
    classical, 15; harmony in, 14–15, 44–45, 110,
    114–15; and language, 96; pattern recogni-
    tion in, 95–97, 106, 110, 124; temporality of,
    95–97, 100–102
musicians, 39, 119–20, 141–42
mu waves, 148–49

Nabokov, Vladimir, 118
Nadal, Marcos, 42
Nagel, Thomas, 6, 113, 177
narrative, 57, 102–3, 144–47, 154–55, 161
narratology, 145–46, 205–6n36, 206n37
natural attitude, 20
Necker cube, 69–71
negation, 87, 169
neural correlates of consciousness (NCC), xii,
    7, 17, 177
neuroaesthetics, xii–xiii, 9–10, 18, 20, 23, 56,
    181–82
neurobiology, ix, xiv, 3–4, 7–8, 18, 21, 177, 179
neurogenesis, 40
neuromodulator hormones, 47, 50
neuronal assemblies, 20, 40, 42, 50, 68, 72, 74,
    79–80, 92, 106–8, 116

neuronal integration, 104, 106, 119, 129,
    146–47; time-scales of, 100–101, 117, 128,
    180. *See also* binding problem; top-down,
    bottom-up interactions
neuronal recycling, x, 2, 27–28, 32, 58, 117–18,
    131, 172, 177. *See also* visual word form area
neurophenomenology, xii–xiii, 19–20, 198n19,
    202n69
neuropsychoanalysis, 185n23
New Criticism, 13, 15, 52, 176
Newtonian causality, 84
New York Yankees, 180
Nicholls, John, 105
Nietzsche, Friedrich: Apollonian and Diony-
    sian, 164; *Gleich-Setzen des Nicht-Gleichen,*
    87, 135, 166
Noë, Alva, 24, 112–13, 126, 184n6
noise, 15, 43, 48, 109–11, 115
novelty, 51, 72, 89, 111, 119–20, 194n22.
    See also constancy; surprise
Nussbaum, Martha, 209n77

olfactory maps, 106
Ong, Walter, 118
open-ended games, 52
opioid receptors, 51
Ortiz, David "Big Papi," 180
out-of-body experience, 156–58
oxytocin, 47, 50, 192n48
Ozawa, Seiji, 128

pain, 149, 160–61, 165
Paradiso, Michael, 104–5
paradox of alter ego, xi, 131, 133–36, 150, 161, 170
parallel processing, x, xiii, 2–3, 12, 27, 41–44,
    58, 107–8
Patel, Aniruddh D., 44, 95, 119
pattern recognition, 14, 23, 57–58, 67, 100–101;
    auditory, 43, 95–97; in reading, 21, 54,
    73–74, 77–79, 110; visual, 58–67
perceiving and processing, 77, 115–16
perceptual framing, 99–100
personal distress, 159
perspectives, 93, 95, 98, 168
Petit, Jean-Luc, 129
Pfeifer, Jennifer, 158
phantom limbs, 157
phenomenology, xii–xiii, 7, 93, 202n69,
    203n77; on intersubjectivity, 133–34; on

reading, 3, 20–22, 84, 87, 118, 147–48; on the self, 127–28; on time, 20, 91–95

phonemes, 32–33, 80–81

photosynthesis, 83

phrenology, 40, 184n6

Pinker, Stephen, 82, 84, 88, 183n3; language instinct, 117, 188n1

pity and fear, 11, 17, 125. *See also* tragedy

plasticity, 2, 14, 26, 28, 40–41, 47–48, 52, 116, 120; constraints on, 32, 34, 36–37, 46

Plato, 169, 178, 189n9, 196n40

play: and ambiguity, 69, 74, 89–90; and art, x, 11, 51–52, 192–93n58; and the brain, ix, 52–53, 86; between cortical areas, 61, 92, 119, 128; in infant imitation, 168–70; in reading, 112, 130, 145, 172; temporality of, 101, 111, 119; to-and-fro of, 52–53, 56, 86

pleasure, 47, 51–52, 112; aesthetic, 37

pluralism, 18

poetic function, 16

point of view, 60. *See also* focalization; perspectives

Pollock, Jackson, 151

portraits, 37

positron emission tomography (PET), 148, 184n6

Poulet, Georges, 135–36, 156–58

Pound, Ezra, "In a Station of the Metro," 85

pragmatic usefulness, 36, 59, 76

prairie voles, 47

Price, Cathy, 33

primordial feelings, 120–22

propaganda, 5

prosopagnosia (face blindness), 38

protentional horizon, 20, 93–94, 96, 101–3, 117, 122–23

psychoanalytic criticism, 85

psychology, 18–19

Pynchon, Thomas, 143; *The Crying of Lot 49*, 46

qualia, 6, 113, 177

quantitative methods, 10

Quiroga, R. Q., 12

Ramachandran, V. S., 1, 13, 41, 133, 148–49, 157, 170, 172, 183n1

ratchet effect, 172

reader-response criticism, xiii, 21–22, 116

readiness, 123–24

reading, xiv, 2, 7–11, 90, 120, 182, 188n1; action words, 122, 153, 160, 165, 202n67; differences between readers, 73–74; as doubling, 135–36, 146, 171; historicity of, 116–18, 178, 179; history of, 27, 117–18, 131, 172; immersion in, 84, 103, 136, 142–48; mirror neurons and, 151, 154–58; neuroscience of, x, 4, 27, 42, 58; pathways to, 33, 67, 81, 166; phenomenology of, 20–23, 54, 84; "real" and "alien" me in, 136, 145–46, 158, 161, 169, 171; reflection in, 143, 145–48; self-other reciprocity in, 142–48, 169–71; as skillful coping, 112, 116, 121; temporality of, 57, 92, 97–98, 100, 110–12, 130

realism, 46, 52, 85, 103, 143, 155

reason vs. feeling, 77

reductionism, 6, 11, 19, 176–77, 179

Reichler, Gerald, 78

relativity. *See* cultural differences; universality

repetition, 111–16, 119

retentional horizon, 20, 93–96, 102–3, 117, 123

retina, 59–62, 64–65

Reynolds, Sir Joshua, 186n29

rhythm: musical, 95–96, 121–22; poetic, 121–22. *See also* brain, rhythms

Richardson, Alan, 187n38

Ricoeur, Paul, 21, 24, 87, 154–55

Riefenstahl, Leni, 5

Rimmon, Shlomith, 195n24

Rizzolatti, Giacomo, xi, 132, 138, 141, 148–49, 157, 163, 170, 208n66

Robbe-Grillet, Alain, 46, 143

rods, 61–62

Roggman, Lori, 38

Romantic aesthetics, 13

Rosch, Eleanor, 99

Roy, Jean-Michel, 126

Russian Formalism, 13

Sacks, Oliver, 38

Sartre, Jean-Paul, 101, 135, 171, 211n105

Saussure, Ferdinand de, 32–33, 63–64, 188n54

schematization, 153–54, 158

Schiller, Friedrich, 51

Schoenberg, Arnold, 45–46, 49, 115

science, 8–9, 11, 175–77, 180–82, 212n10

scientific fact, 8, 184–85n18

sculpture, 164

Searle, John, 128–29, 177, 179–80

sea slug (*Aplysia californica*), 113–15
self, the, 126–27
self-consciousness, 89–90, 102
self-other reciprocity, 169–71
sensitization, 113–15
sensory modalities, 164, 166
Seymour, Philip, 34
Shakespeare, William, 18; *Hamlet*, 85
shallow orthographies, 33
shared manifold, 205n25
Shklovsky, Victor, 13, 48, 111
Sidney, Sir Philip, 169
signs, 27–28, 33; graphic, 30, 32; linguistic, 63–64, 178, 188n54; visual, 31
simulation, 125–26, 146–47, 160, 163–64, 169, 172; embodied, 151; high- and low-level, 144; theory (ST), xi, 132, 134, 142–43, 145, 168
simultaneity, 98–102, 108
Singer, Tania, 146, 159–61
skin, 158, 208n67
Skov, Martin, 6, 47
sleep, 108–10
Slingerland, Edward, 185n22
smell, 106, 124, 162–63, 166
Snow, C. P., 175
social constructionism, 8, 184–85n18
social emotions, 121
social sciences, 10
solipsism, xi, 133–35
somatic markers, 76–77, 195–96n31
speech, 152–53
Spolsky, Ellen, 64, 191n35
spoonerisms, 81
Stafford, Barbara Maria, 197n45
staircase illusion, 75–76
Starr, G. Gabrielle, 121, 124, 185n23
Stewart, C. Matthew, 40
strata of literary work, 22
Stravinsky, Igor, 115; *Rite of Spring*, 49
subliminal priming, 123–24
surprise, xi, 23, 43–44, 56–57, 85–87, 94, 98, 121, 124, 145, 171
symbolic action, 154–56, 165
synapse, 99, 103–4
synchrony, 105–10, 124

Tallis, Raymond, 183n5
taste, 124
teaching, 9, 144, 177

temporality: of decentered self, 126–30; emotions and, 121–26; of interpretation, 56–57, 81, 97–103, 122–25; and music, 96–97, 110–11; of neuronal assemblies, xi, 20, 23, 98–102, 106–10, 119–20; phenomenological paradoxes of, 91–95; of reading, 97–98, 102–3, 110–13, 116–20, 154
terministic screens, 7–8
Tervaniemi, Mari, 43–44
textual indeterminacies, 21–22
theory of mind (ToM), xi, 18, 132, 134, 141–43, 145–47, 168
Thompson, Evan, 94, 97, 99, 101, 106, 112, 120, 131, 202n69, 204n13
Tintoretto, 156, 158
to-and-fro movement: in cortical processes, 59–60, 67, 86, 111–12, 119, 146; in interpretation, 55–56, 82–90, 92, 97; in play, 51–52, 129–30, 169–72; in reading, 84–89, 92, 117, 154, 170–71
Tolstoy, Leo: *Anna Karenina*, 144, 198n16; *The Kreutzer Sonata*, 209n88
Tomasello, Michael, 171
top-down, bottom-up interactions: of cortical regions, 27, 77, 119–20, 127–29, 141; and multi-stable images, 67, 69–73; in reading, 33–34, 40–44, 79–80; and temporality of the brain, 20, 23, 40, 100, 104–5; in vision, 58–62, 67
touch, 58, 166–67
tragedy, 155, 164–65. *See also* pity and fear
transcranial magnetic simulation (TMS), 151–53
trauma, 201n62
trees, branching and recursive, 82–83
Trehub, Sandra, 45
Turner, Mark, 87–88
turn-taking, 170, 172
two cultures, 10, 175–77

universal grammar, 188n1
universality, 2, 14–16, 32, 37–38, 44, 120–21, 178
unwritten vs. written text, 84

validity, 8–9, 57, 179
value, 22, 43, 50
Van Gelder, Tim, 95–96
Varela, Francisco J., xi, 19–20, 41, 94, 98–99, 106–7, 117, 123
Vartanian, Oshin, 43, 47

vasopressin, 47, 50, 192n48
Vermeule, Blakey, 144
Vessel, Edward A., 41
Victorian novel, 119
violence: in movies, 125, 172–73; in video
  games, 173–74
virtual dimension, 84–86
vision, 4, 15, 26, 35–36, 41, 188n1; ambiguity
  and, 67–72; hermeneutic processes of, 30,
  32, 58–67, 81, 84, 166–67; neurobiology of,
  8, 11, 35–36, 41
visual arts, 26, 85, 164–65
visual word form area (VWFA), 4, 28–29,
  32–33, 36, 42, 78–80, 116, 119, 147, 189n6
Vivaldi, Antonio, "Four Seasons," 49
voxel, 4, 206n44

Wallace, David Foster, 46
"What" and "Where" systems, 62

Wicker, Bruno, 162–63
Wilde, Oscar, 147
Willems, Roel W., 202n67
Wilson, Edward O., 179, 212n10
Wittgenstein, Ludwig, 68, 144
Woolf, Virginia, 17, 112
word recognition, 21, 26, 28, 30, 32–33, 78–79
writing, 28, 30–31, 33, 153

Zahavi, Dan, 91, 93, 115, 123–24, 126, 150
Zaidel, Dahlia, 43
Zbikowski, Lawrence M., 96
Zeki, Semir, 6, 47; on aesthetics, 37, 42, 84–85,
  89, 197n46; on ambiguity, xi, 55–56, 67–72,
  75; on disunity of consciousness, 77, 99–100,
  128–29, 199n22; on vision, 35, 59, 63–64
Zeman, Adam, 6
Zimbardo, Philip, 173
Zunshine, Lisa, 147